Africa's ICT Infrastructure

Africa's ICT Infrastructure

Building on the Mobile Revolution

Mark D. J. Williams, Rebecca Mayer, and Michael Minges

Vivien Foster and Cecilia Briceño-Garmendia
Series Editors

THE WORLD BANK
Washington, D.C.

About the AICD

This study is a product of the Africa Infrastructure Country Diagnostic (AICD), a project designed to expand the world's knowledge of physical infrastructure in Africa. The AICD provides a baseline against which future improvements in infrastructure services can be measured, making it possible to monitor the results achieved from donor support. It also offers a more solid empirical foundation for prioritizing investments and designing policy reforms in the infrastructure sectors in Africa.

The AICD was based on an unprecedented effort to collect detailed economic and technical data on the infrastructure sectors in Africa. The project produced a series of original reports on public expenditure, spending needs, and sector performance in each of the main infrastructure sectors, including energy, information and communication technologies, irrigation, transport, and water and sanitation. The most significant findings were synthesized in a flagship report titled *Africa's Infrastructure: A Time for Transformation*. All the underlying data and models are available to the public through a Web portal (http://www.infrastructure africa.org), allowing users to download customized data reports and perform various simulation exercises.

The AICD was commissioned by the Infrastructure Consortium for Africa following the 2005 G-8 Summit at Gleneagles, which flagged the importance of scaling up donor finance to infrastructure in support of Africa's development.

The first phase of the AICD focused on 24 countries that together account for 85 percent of the gross domestic product, population, and infrastructure aid flows of Sub-Saharan Africa. The countries were Benin, Burkina Faso, Cape Verde, Cameroon, Chad, Democratic Republic of Congo, Côte d'Ivoire, Ethiopia, Ghana, Kenya, Lesotho, Madagascar, Malawi,

and efficiency, and estimates of the need for additional spending on investment, operations, and maintenance. Each volume also comes with a detailed data appendix—providing easy access to all the relevant infrastructure indicators at the country level—which is a resource in and of itself.

In addition to these sector volumes, the AICD has produced a series of country reports that weave together all the findings relevant to one particular country to provide an integral picture of the infrastructure situation at the national level. Yet another set of reports provides an overall picture of the state of regional integration of infrastructure networks for each of the major regional economic communities of Sub-Saharan Africa. All of these papers are available through the project Web portal, http://www.infrastructureafrica.org, or through the World Bank's Policy Research Working Paper series.

With the completion of this full range of analytical products, we hope to place the findings of the AICD effort at the fingertips of all interested policy makers, development partners, and infrastructure practitioners.

Vivien Foster and Cecilia Briceño-Garmendia

About the Authors

Mark D. J. Williams is a Senior Economist in the Global ICT (information and communication technologies) group of the World Bank. He has carried out extensive research into the economics and regulation of the ICT sector and has been involved in the development of World Bank policy on broadband and on public private partnerships in the sector. Mr. Williams has worked on the design and implementation of ICT infrastructure projects around the world and has advised governments and operators on a wide range of policy and regulatory issues. He regularly publishes articles and speaks at conferences on the sector, recently focusing on Africa and the Middle East. Mr. Williams has degrees from Oxford and Warwick universities in the United Kingdom.

Rebecca Mayer is a consultant on telecommunications and energy infrastructure in Africa and other developing regions. She specializes in techno-economic modeling of wireless telecommunications and distributed power solutions, as well as the use of ICTs in off-grid and energy-constrained environments. In addition to her work for the World Bank, Ms. Mayer has previously served as a Program Officer in the Clean Energy Group at Winrock International, as a Research Officer at the International Telecommunication Union, and as a Senior Associate at Pyramid Research/Economist Intelligence Unit. Ms. Mayer holds master's degrees in Engineering and Public Policy from Carnegie Mellon University and in Economics from Tufts University, as well as a Bachelor of Arts in Near Eastern Studies from Yale University.

Michael Minges is an international ICT consultant. Mr. Minges was formerly head of the market research unit at the International Telecommunication Union, where his functions included the analysis of telecommunications trends in developing nations, market studies of countries and regions, and regulatory and policy advice. He also worked at the International Monetary Fund as an information technology specialist. Mr. Minges regularly advises governments and private clients on key ICT policy, strategy, and regulatory issues. He has written numerous articles and makes frequent presentations around the world on telecommunications progress in emerging countries. He launched a principal industry publication, the *World Telecommunication Development Report* and designed a leading indicator for measuring ICT progress, the Digital Access Index. Mr. Minges holds a Bachelor of Arts degree in Economics from the University of Oregon and a Master of Business Administration degree from George Washington University.

Acknowledgments

This book was coauthored by Mark D. J. Williams, Rebecca Mayer, and Michael Minges, under the overall guidance of series editors Vivien Foster and Cecilia Briceño-Garmendia.

The book draws on background papers that were prepared by World Bank staff and consultants, under the auspices of the Africa Infrastructure Country Diagnostic (AICD). Key contributors to the book on a chapter-by-chapter basis were as follows.

Chapter 1

This chapter is an introduction to the work, written by Mark D. J. Williams.

Chapters 2 and 3

Contributors

Mavis Ampah, Cecilia Briceño-Garmendia, Daniel Camos, Michael Minges, Rupa Ranganathan, Maria Shkaratan, and Mark D. J. Williams

Key Source Document*

Ampah, M., D. Camos, C. Briceño-Garmendia, M. Minges, M. Shkaratan, and M. Williams. 2009. "Information and Communications Technology in Sub-Saharan Africa: A Sector Review." AICD Background Paper 10, World Bank, Washington, DC.

Chapter 4

Contributors

Mark D. J. Williams, Jacqueline Irving, and Astrid Manroth

Key Source Document*

Irving, J., and A. Manroth. 2009. "Local Sources of Financing for Infrastructure in Africa: A Cross-Country Analysis." Policy Research Working Paper 4878, World Bank, Washington, DC.

Chapter 5

Contributors

Rebecca Mayer, Ken Figueredo, Mike Jensen, Tim Kelly, Richard Green, Alvaro Federico Barra, and Mark D. J. Williams

Key Source Document*

Mayer, R., K. Figueredo, M. Jensen, T. Kelly, R. Green, and A. F. Barra. 2009. "Connecting the Continent: Costing the Needs for Spending on ICT Infrastructure in Africa." AICD Background Paper 3, World Bank, Washington, DC.

Chapter 6

This chapter was written by Mark D. J. Williams. It contains a synthesis of the analysis presented in the other chapters and presents policy recommendations intended to drive the sector forward.

None of this research would have been possible without the generous collaboration of government officials in the key sector institutions of each country, as well as the arduous work of local consultants who assembled this information in a standardized format.

The work benefited from widespread contributions and peer review from colleagues within the World Bank, notably Mavis Ampah, Laurent Besançon, Jérôme Bezzina, Doyle Gallegos, Tim Kelly, Ioannis Kessides, Kaoru Kimura, Juan Navas-Sabater, Rupa Ranganathan, Carlo Rossotto, Lara Srivastava, and Eloy Eduardo Vidal. The external peer reviewers for this volume, Charley Lewis and Russel Southwood, provided constructive and thoughtful peer review comments. The comprehensive editorial effort of Steven Kennedy is much appreciated.

---

*All source documents are available for download from the AICD website, http://www.infrastructureafrica.org.

Abbreviations

Costs are expressed in U.S. dollars unless stated otherwise.

3G	third generation
AAA	authentication, authorization, and accounting
ACE	Africa Coast to Europe (cable)
ADSL	asymmetric digital subscriber line
AfDB	African Development Bank
AICD	Africa Infrastructure Country Diagnostic
AIKP	Africa Infrastructure Knowledge Program
AIX	Accra Internet Exchange
ARPU	average revenue per user
ASB	Alcatel Shanghai Bell
BCIE	Banco Centroamericano de Integracion Economica
BESA	Bond Exchange of South Africa
BOAD	Banque Ouest Africaine de Développement
BRVM	Bourse Régionale des Valeurs Mobilières
BSC	base station controller
BTS	base transceiver station
CAPEX	capital expenditure
CCT	China-Congo Telecom

CDMA	Code Division Multiple Access
CEMAC	Economic and Monetary Community of Central Africa
CIAT	Centro Internacional de Agricultura Tropical
CIESIN	Center for International Earth Science Information Network
COMESA	Common Market for Eastern and Southern Africa
CPE	customer premises equipment
CST	Companhia Santomense de Telecomunicacaoes
DBSA	Development Bank of South Africa
DEG	Deutsche Investitions- und Entwicklungsgesellschaft mbH
DFI	development finance institution
DFID	Department for International Development
DGE/DGI	Direction des Grandes Entreprises/Direction Générale des Impôts
DGRAD	Direction Générale des Recettes Administratives, Judiciaires, Domaniales et de la Participation
DSL	digital subscriber line
EAC	East African Community
EASSy	Eastern African Submarine Cable System
ECCAS	Economic Community of Central African States
ECOWAS	Economic Community of West African States
EIB	European Investment Bank
EIG	Europe-India Gateway
ESCOM	Electricity Supply Corporation of Malawi
EVDO	Evolution–Data Optimized
FALCON	FLAG Alcatel-Lucent Optical Network
FLAG	Fiber-Optic Link around the Globe
FMO	Nederlandse Financierings-Maatschappij voor Ontwikkelingslanden N.V. (Netherlands Development Finance Company)
GAMTEL	Gambia Telecommunication Company Ltd.
Gbps	gigabits per second
GDP	gross domestic product
GGSN	gateway GPRS support node
GHz	gigahertz
GIX	Ghana Internet Exchange
GLO-1	Globacom-1
GNI	gross national income

GPRS	general packet radio service
GRUMP	Global Rural-Urban Mapping Project
GSM	global system for mobile communications
GSMA	GSM Association
HA	home agent
HH	household
HHI	Herfindahl-Hirschman Index
HLR	home location register
HSPA	HighSpeed Packet Access
ICASA	Independent Communications Authority of South Africa
ICT	information and communication technology
IDA	Infocomm Development Authority
IFC	International Finance Corporation
IPLC	international private lease circuit
ISM	industrial, scientific, and medical (radio band)
ISP	Internet service provider
ITU	International Telecommunication Union
IXP	Internet exchange point
JSE	Johannesburg Securities Exchange
Kbps	kilobits per second
KfW	Kreditanstalt Für Wiederaufbau
KIXP	Kenya Internet Exchange Point
km	kilometer
km^2	square kilometer
kWh	kilowatt-hour
LAP	Libyan Arab Portfolio
LIBOR	London Interbank Offered Rate
LION	Lower Indian Ocean Network
LTE	Long Term Evolution
MAURITEL	Société Mauritanienne de Telecommunications
Mbps	megabits per second
MCEL	Moçambique Celullar
MHz	megahertz
MNP	mobile number portability
MSC	mobile switching center
MTC	Mobile Telecommunications Company
MTL	Malawi Telecom Ltd.
MTR	mobile termination rate
MVNO	mobile virtual network operator

NB&D	network, build, and deploy
NEPAD	New Partnership for Africa's Development
NITEL	Nigerian Telecommunications Ltd.
NRA	national regulatory agency
NRPT	National Rural Telephony Project
ODA	official development assistance
OECD	Organisation for Economic Co-operation and Development
ONATEL	Office National des Télécommunications
OPEC	Organization of the Petroleum Exporting Countries
OPEX	operating expenditure
PDA	personal digital assistant
PDSN	packet data serving node
PPI	private participation in infrastructure
PPP	purchasing power parity
PV	present value
RCDF	Rural Communications Development Fund
RIO	reference interconnection offer
SADC	Southern African Development Community
SAFE	South Africa Far East
SAT-1	South Atlantic 1 cable
SAT-2	South Atlantic 2 cable
SAT-3/WASC	South Atlantic 3/West Africa Submarine Cable
SE	stock exchange
SEA-ME-WE	South East Asia–Middle East–West Europe
SGSN	serving GPRS support node
SIM	subscriber identity module
SMS	short message service
SOCATEL	Société Centrafricaine de Télécommunications
SOE	state-owned enterprise
SONATEL	Société Nationale des Télécommunications du Senegal
SONITEL	Niger Telecommunications Co.
SOTELGUI	Société des Télécommunications de Guinée
SOTELMA	Société des Télécommunications du Mali
SRTM	shuttle radar topography mission
Tbps	terabits per second
TCRA	Tanzania Communications Regulatory Authority
TDB	Telecommunication Development Bureau
TdM	Telecomunições de Moçambique

TDMA	Time Division Multiple Access
TEAMS	The East African Marine System
TELMA	Telecom Malagasy
THL	Telecom Holdings Ltd.
TRX	transceiver
TTCL	Tanzania Telecommunications Company Ltd.
UCC	Uganda Communications Commission
UEMOA	Union Économique et Monetaire Ouest-Africaine
UMTS	Universal Mobile Telecommunications System
UN	United Nations
USE	Uganda Securities Exchange
VAT	value added tax
VLR	visitor location register
VoIP	Voice Over Internet Protocol
VRA	Volta River Authority
WACS	West Africa Cable System
WAEMU	West African Economic and Monetary Union
WCDMA	Wideband Code Division Multiple Access
WiMAX	Worldwide Interoperability for Microwave Access
WLL	wireless local loop
Zamtel	Zambia Telecommunications Company
ZCC	Zambia Competition Commission
ZESCO	Zambia Electricity Supply Corporation Ltd.
ZTE	Zhong Xing Telecommunication Equipment Company Ltd.

CHAPTER 1

Introduction

Information and communication technologies (ICTs) have been a remarkable success in Africa. Across the continent, the availability and quality of service have gone up and the cost has gone down. In just 10 years—dating from the end of the 1990s—mobile network coverage rose from 16 percent to 90 percent of the urban population; by 2009, rural coverage stood at just under 50 percent of the population. Institutional reform has driven this radical change in telecommunications. Markets have been liberalized, and regulatory bodies have been established. The resulting increase in competition has spurred investment and dramatic reductions in prices.

The speed at which the sector has evolved, the nature of the policy changes that have triggered the reforms, and the way in which investment has been financed all make telecommunications unique among the infrastructure sectors in Africa. Despite the successes of recent years, however, several major challenges remain for policy makers.

The first of these challenges is to continue the expansion of the mobile networks, bringing basic voice services to as much of the population as possible. To do this, policy makers need answers to key questions: What have been the drivers of past expansion? Why do some countries in the region consistently outperform others? How far will the current model of reform go toward providing universal coverage? The answers to these questions can serve as a foundation for designing policy that fosters the success of the mobile voice revolution across the region.

Although the performance of Africa's mobile networks over the past decade has been remarkable, the telecommunications sector in the rest of the world has also evolved rapidly. Many countries now regard broadband Internet as central to their long-term economic development strategies, and many companies realize that the use of ICT is the key to maintaining profitability. In Africa, however, the Internet is still in its infancy. In most countries, access is limited and slow. Where broadband is available, it is typically very expensive—far beyond the financial means of the majority of Africans. Ensuring that networks are capable of delivering broadband Internet access at affordable prices is the next major challenge on the horizon for policy makers.

This book is about that challenge and others. Chapters 2 and 3 describe the recent history of the telecommunications market in Africa; they cover such issues as prices, access, the performance of the networks, and the regulatory reforms that have triggered much of the investment. This part of the book compares network performance across the region and tries to explain why some countries have moved so much more quickly than others in providing affordable telecommunications services.

Chapter 4 explores the financial side of the telecommunications revolution in Africa and details how the massive investments have been financed and which companies have most influenced the sector.

Chapter 5 deals with the future of the sector, addressing some of the main policy questions that it faces: How far will the expansion of mobile voice networks go under the current policy regime? How much of the population is likely to be living outside the region's commercially viable zones? Is it commercially viable to provide broadband Internet to broad segments of the population, in addition to large businesses and high-income individuals? Is there any way in which broadband Internet will develop into a mass-market service in Africa?

The final chapter synthesizes the main chapters of the book and presents policy recommendations intended to drive the sector forward.

Access to Communications

Since the end of the 1990s, the availability of telecommunications services has dramatically increased across Sub-Saharan Africa. Networks have expanded and prices have fallen, bringing basic telecommunications services within reach of the majority of Africans. This success, however, does not extend to all segments of the market. The number of subscribers to fixed-line networks, relative to the size of the population, has been static, and although more and more Africans are accessing the Internet, online use in Sub-Saharan Africa still lags far behind that in other parts of the world.

Access

The explosion in access to telecommunications services has been most prominent in the mobile market, in which the number of users grew by more than 247 million between 1998 and 2008. The mobile penetration rate increased from less than 1 percent of the population in 1998 to almost one-third by 2008—and since then has continued to increase.[1] Moreover, rapid growth has occurred throughout the region. Low-income countries, where telecommunications services were once accessible only to a privileged few, are quickly catching up with their richer neighbors, such as Namibia and South Africa. In 1998, at the start of Africa's telecommunications revolution, South Africa accounted for 86 percent of all subscribers in the region, but by 2008, that figure was down to 18 percent; Nigeria overtook South Africa as the region's biggest telecommunications market in 2008.

Other segments of the telecommunications market have not developed nearly as quickly as the mobile businesses. The number of fixed lines, for example, increased from 1.4 subscribers per 100 people in 2000 to 1.5 in 2007 before falling to 1.4 in 2008. Internet access across Africa is also very low: Penetration rates on the continent are a fraction of those in other regions, but these rates are slowly increasing as wireless broadband technology becomes more established and prices fall.

Despite the overall improvement in access to telecommunications in Africa, some countries have been much more successful than others. Nearly seven times as many mobile subscribers per capita are found in the most successful 10 countries as in the least successful 10, despite global standards and broadly similar costs and operating conditions across the region.

Prices

High prices have historically been one of the factors preventing the public from accessing ICT services in Africa. Although the price of specific telephone services (such as local calls, international calls, and text messages) varies, a standard basket of services that includes subscription this distinction really applies only to the "last mile" (that is, the link between the network and the customer, also known as the "access network"), because much of a mobile network is actually composed of fixed (usually fiber-based) links. The idea of physically separate networks is also not an accurate picture of the way modern telecommunications networks are designed. Operators buy and sell network services to one another, in effect using other operators' networks to fill in gaps in their own infrastructure. For example, mobile operators in many high-income countries may own only the wireless "last mile" infrastructure, while relying on other operators for other parts of their network.

In Africa, the network architecture is different. The integration of networks is less advanced; networks have traditionally been built as standalone, end-to-end networks. Mobile networks are more likely to be wireless throughout, rather than fiber optic at their core and fixed wireless technologies are often used to provide the last-mile links to businesses and homes in place of copper.

These technology-based distinctions are starting to become obsolete. Fixed wireless networks are becoming mobile, wireless networks are being upgraded to fiber, and networks that were once used to provide voice services are increasingly being used to provide a full range of ICT services. For now, however, traditional concepts of fixed and mobile, and wireline and wireless, remain useful analytically and are used throughout this volume.

Fixed-line copper-based connections were the traditional means of linking customers to the telephone network. In Africa, these networks have always had very low penetration levels, and in many cases, the levels have fallen over time. Mobile networks have provided a ready substitute for these traditional networks for basic voice services while offering added mobility, lower costs, and more payment flexibility.

The rapid growth in mobile network infrastructure, however, has greatly expanded access to telecommunications in Africa. Networks that were initially concentrated in towns and cities increasingly began pushing into rural areas. By 2009, 90 percent of Africa's urban population and 48 percent of its rural population lived within reach of a mobile network.[2] Coverage numbers continue to increase, although signs indicate that the rate of increase is slowing as networks expand into less economically viable areas.

As for offering access to broadband, Africa has followed a very different infrastructure growth path than high-income countries. The dominant form of broadband Internet-access infrastructure in the OECD countries has been copper based, either upgraded telephone lines or upgraded cable TV networks. Wireless broadband in these countries has generally been seen as a complement to wireline access rather than a substitute for it. In Africa, by contrast, the lack of suitable copper wireline infrastructure has not only limited access to broadband Internet but also increased the role of wireless network infrastructure in providing such access. Many global wireless broadband standards are used, including the third-generation family of mobile standards (3G),[3] Worldwide Interoperability for Microwave Access (WiMAX), and Long Term Evolution (LTE). WiMAX was the first widely deployed broadband access network infrastructure in Africa. More recently, mobile operators have begun upgrading their networks to be able to provide 3G, and some are using LTE on a trial basis. These technologies are expected to play an increasingly important role in delivering broadband to customers in Africa, because few signs exist that wireline access infrastructure is going to develop in a significant way for the foreseeable future.

As the number of subscribers, particularly broadband Internet users, increases, traffic levels on the networks grow. Although the access, or last-mile, networks are likely to focus on wireless technologies, operators are increasingly upgrading their core, or "backbone," networks to fiber-optic technologies. These high-capacity networks lie at the heart of any modern broadband communications system, even if the number of subscribers is relatively limited. Although much of the investment in mobile networks has gone into wireless infrastructure, the fiber-optic networks in Africa are developing quickly. As of the end of December 2009, the operational terrestrial fiber-optic transmission network in Sub-Saharan Africa was 234,000 kilometers (km) long, with a further 41,000 km under construction.

Historically, the key players in the development of fiber-optic networks have been state-owned telecommunications operators. This is gradually changing as new private operators enter the fiber-optic network business. Private operators are building about half the length of fiber-optic cables currently under construction; this share is expected to increase as more of the mobile companies move into the data business. Private investment in fiber is dependent on a conducive regulatory framework. Countries that have encouraged investment in fiber networks through issuing licenses and assisting operators to obtain rights of way, for example, have achieved much better results than those that have placed restrictions on fiber networks. Such restrictions include licensing constraints that limit the potential size of operators' markets, obstacles to obtaining rights of way, and, in some cases, outright monopolies.

In countries that have encouraged investment in fiber-optic networks, development has followed a similar pattern. The most commercially viable routes for fiber-optic upgrades run along corridors connecting major urban areas with one another and with exit points from the country (such as submarine-cable landing stations and border crossings). Fiber-optic cables cluster along these routes, and networks compete directly with one another. This pattern has positive effects: The competition tends to push down prices, and the multiple networks provide resilience. At the same time, route concentration leaves much of the region—especially rural areas—unconnected to the backbone networks. Providing broadband access to these areas will be a key policy challenge for the sector.

Two recent trends in the sector have implications for further development of domestic fiber-optic networks. The first is the emergence of regional wholesale carrier network operators that provide cross-border services to corporate customers and other licensed operators. Second, fixed-line operators and mobile operators in contiguous neighboring countries have in some cases fallen under common ownership through privatization, acquisition, and new start-ups. The owners then use their licenses in multiple countries to interconnect their networks. These trends are driving the development of long-haul infrastructure that is connecting countries within subregions and linking landlocked countries to the landing points of the submarine cables. Competition between these networks will lower prices and increase the capacity available to users, giving access to submarine cable infrastructure for the landlocked countries. But this process is more advanced in some parts of Africa than others. Within East Africa, for example, long-haul fiber-optic networks are being rolled out quickly. In Central and West Africa, however, cross-border network development is still at an early stage.

Broadband generates higher volumes of traffic than voice services do. Furthermore, a greater proportion of this traffic is international because much of the content that is accessed over the Internet is stored in countries other than the one in which the user is located. This is particularly true of Africa, which currently hosts very little Internet content. The capacity of international networks has therefore been a key constraint on the development of affordable broadband services in Africa.

Voice networks in Africa have traditionally relied on satellites to handle international traffic because of the lack of submarine fiber-optic network infrastructure and the relatively low bandwidth requirements of international voice traffic. Recently, however, submarine fiber-optic cable projects have proliferated. As of 2010, 12 submarine cables were operational in Sub-Saharan Africa and five more were being deployed. The operational cables have a combined capacity of over 12 terabits per second (Tbps). A total of $1.7 billion is being invested in the five submarine cables currently under construction, which will bring an additional 9 Tbps of capacity to the region. Clearly, the amount of international bandwidth available to Africa is growing rapidly. It therefore looks likely that what was once a major infrastructure bottleneck in Africa will no longer be a constraint on the sector in the future.

Institutions and Market Reform

Major reform in sector management has been the key factor in the dramatic improvement in ICT services in Africa. Previously, a state-owned monopoly operator provided all ICT services in a country. Beginning in the 1990s, however, African governments began liberalizing their telecommunications markets by issuing multiple licenses and allowing operators to compete with one another. Private investment now drives network expansion, and privately owned operators are the main service providers. This shift has been accompanied by reform of the government institutions that are responsible for the sector. Regulatory authorities have been established, and in many cases the formerly state-owned operators have been privatized.

Market Liberalization

The widespread liberalization of markets and the emergence of competition have greatly increased the performance of Africa's ICT sector. The growth in competition among mobile operators has been particularly rapid. Most countries now have multiple mobile operators that compete with one another. By comparison, in the 1990s mobile networks were usually monopolies—if they existed at all (figure 1.1).

The increase in mobile competition has been accompanied by growth in the number of subscribers, and that growth has accelerated as competition has intensified. Growth in subscriber numbers has generally been modest following the initial stage of market liberalization—that is, the move from monopoly to duopoly. Once a country issues its fourth mobile license, however, penetration rates increase by an average of about 4 percentage points per year.

Despite the clear benefits of market liberalization, some countries have not moved as quickly as others. Although few countries in Africa retain legal prohibitions on competition in telecommunications, other

Figure 1.1 Competition in Mobile Markets in Sub-Saharan Africa, 1993–2009
percentage of countries with no provider, one provider, two providers, and three or more providers

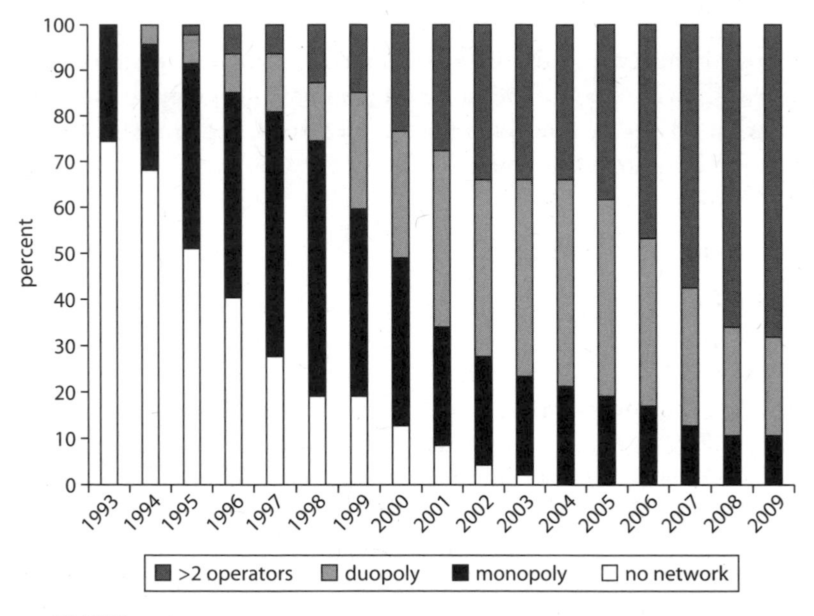

Sources: ITU (2010), regulators, operators.

legal or institutional factors may prevent competition from developing in all segments of the market. For example, existing operators often have exclusivity clauses in their licenses or other types of commitment from the government to issue no further licenses for a defined period of time. Competition in the fixed segment of the market also often lags behind that in the mobile segment. Although 16 Sub-Saharan African countries had fixed-line competition by 2009, only a few of them had more than two operators, for a combination of regulatory and economic reasons. Some countries that have formally ended a fixed-line monopoly have taken time to issue a second fixed-line license. In other cases, fixed-line licenses have been available, but only limited interest from investors has been seen.

Overall, the state of market liberalization across the region is best described as incomplete. Although most countries have multiple mobile operators, few have more than three, despite evidence that most markets in the region can support more. Meanwhile, the process of liberalizing fixed-line markets is not far along. This is despite the fact that countries, such as Nigeria, have shown that a competitive fixed-line market is possible and can deliver benefits to customers in terms of both price and services.

Private Sector Participation

Market liberalization and regulatory reform have triggered large-scale investment in telecommunications networks, mainly from the private sector. Most countries have adopted a relatively liberal policy toward foreign investment in telecommunications, allowing foreigners to own at least 51 percent of telecommunications companies, and some have gone even further by allowing foreign investors to have complete ownership of subsidiaries. Only four countries (the Comoros, Djibouti, Eritrea, and Ethiopia) retain major restrictions on foreign investment in operators.

In general, private investors have favored investment in greenfield projects (that is, the purchase of a license without any existing business or network assets) over privatization. In fact, greenfield projects have attracted three-quarters of all investment in physical assets in the sector, with most of the rest going to privatized incumbent operators. Most of these projects have involved mobile operators, although fixed-line operators have also proved attractive to investors in some countries, such as Nigeria and Sudan.

Where does private investment in Africa's telecommunications sector come from? One might expect Europe or the United States—the pioneers of mobile telephone networks—to lead the drive to invest in similar businesses in Africa. Yet this has not been the case. There was an initial flurry of investment by European and American operators in privatized incumbents in the late 1990s. Since then, however, most investment in the telecommunications sector—for both greenfield projects and privatization—has come from investors in other developing countries. For example, Morocco-based Maroc Telecom has participated in the privatization of incumbent operators in Burkina Faso, Gabon, Mali, and Mauritania, and Malawi Telecom Ltd. (MTL) was sold to a group of mainly domestic investors. In all, more than 80 percent of the investment in private operators in Sub-Saharan Africa has come from companies based in Africa or the Middle East. Only recently have investors from developed countries re-entered Africa's telecommunications market in a significant way: France Telecom bought 51 percent of Telkom Kenya in 2007, and Vodafone, based in the United Kingdom, bought 70 percent of Ghana Telecom in 2008.

Regulation

The shift from monopoly to competition in the ICT sector has been accompanied by reform of the legal framework that governs it. All African countries have introduced new laws and regulations covering telecommunications. Typically, the new laws and associated regulations establish a national regulatory agency (NRA) along with general provisions for competition, licensing, interconnection, managing scarce resources (particularly the radio spectrum), and pricing. The primary role of an NRA is to establish and implement the rules that govern the sector and to protect customers by regulating prices and monitoring the quality of service. The NRA's decisions affect the pace at which competition intensifies, the amount that operators invest, and the extent to which customers benefit from improved services and lower prices. By 2009, 41 African countries had established independent NRAs, up from 5 in 1996.

The effectiveness of an NRA depends on many factors, including the nature of the decision-making process, how the institution is financed, how senior staff members are selected, and the terms of their employment. Operational independence from government helps the NRA to make unbiased decisions without undue political influence and is therefore essential to its effectiveness. The way in which the authority is financed is a key aspect of this independence. Most regulatory authorities are financed through license fees, providing them some autonomy from the central government. Budgets often have to be approved by the government or parliament, however, so some residual political influence over regulatory activities remains.

Another important aspect of operational independence is the way in which the senior staff of the regulatory authority are appointed. Among the heads of regulatory bodies in Sub-Saharan Africa, 78 percent are appointed by either heads of state, the parliament, or a council of ministers. This method of appointment contributes to regulatory independence from the day-to-day politics of the sector. In other countries, however, the sector minister retains the power to appoint the head of the regulatory authority, often resulting in increased political influence over regulatory decisions.

Regulators have a wide range of responsibilities, including licensing, arbitrating disputes, setting tariffs, monitoring the sector, and implementing the universal service policy. These responsibilities have evolved as competition has developed: Whereas regulators used to focus mainly on controlling the tariffs of the incumbent operator, now they must regulate an increasingly competitive market. Most countries have followed a broadly similar path of liberalization in the sector, so regulators are facing similar challenges across the region. One of the most significant of these is regulating the terms on which operators are interconnected. This is a crucial factor in the development of competition and the extent to which benefits are passed on to customers. African regulators are gradually moving toward a system that bases interconnection tariffs on costs. One of the first decisions in this area occurred in Botswana in 2003, where the regulator resolved a dispute between operators by imposing a rate based on a benchmark of tariffs from European operators. This was followed in 2004 by a decision by the Tanzanian regulatory authority imposing mobile interconnection tariffs that were based on costs. Since then, other regulators have followed suit.

Another key role of regulators is to design and implement universal service funds, which have been established in many countries to provide services to underserved areas of the country. They are usually financed through a levy on the sector, which the regulator is typically responsible for collecting. The amount of discretion that the regulator has in determining how the funds are spent varies among countries. The regulator often has an executive role and is responsible for spending the funds once they are allocated.

The third major area of responsibility for telecommunications regulators is the management of the radio spectrum. Across Africa, the organization and management of the spectrum is done in the traditional way—public authorities decide what the different radio-spectrum bands will be used for, how they will be allocated, and how much they will cost to use. Regulators are often then given responsibility for implementing the governments' decisions. The lack of widespread wireline infrastructure in Africa means that the success of the telecommunications sector is even more dependent on the efficient management of the radio spectrum than in countries where many telecommunications services are provided over copper or fiber-optic cable. Globally, a trend is seen toward using more flexible and market-based methods for allocating and managing the radio spectrum. Initial allocations of the radio spectrum are often done by auction, and then, importantly, the allocations are transferable between private parties. This reduces the direct control of public authorities over the radio spectrum and allows market forces to have a greater influence on how it is used. Such mechanisms have yet to be introduced in Africa, but given the dependence of the sector on the radio spectrum, they could have a significant positive impact.

Widespread regulatory reform has been a major factor in the success of the ICT sector in Africa. The gradual shift from a politically driven decision-making process to a more rules-based, technocratic one has improved investor confidence and allowed competition to develop. Independent regulatory authorities were relatively unknown in Africa before the establishment of telecommunications regulators. Although it has taken time for these institutions to become effective, their technical capacity and experience in carrying out their mandate has improved, and the quality of regulation has increased.

Reform and Performance

The dramatic changes in the way that countries manage their telecommunications sectors have had an impact not only on investment, but also on the way that services are delivered and the efficiency with which the sector is managed. Mounting evidence suggests that as markets become more competitive, performance improves, which, in turn, stimulates greater levels of investment, more extensive networks, and lower prices.

It takes time before the market outcomes of policy reforms can be seen. Markets in countries that were the earliest to reform and that went furthest in establishing effective competition have therefore exhibited better performance than countries that delayed the process. For example, over the past 10 years early reformers have experienced higher mobile penetration rates than countries that reformed later or less thoroughly. Sector performance has been very poor in the few countries in Africa that have not introduced any major structural reforms. In the mobile segment of the market, increased competition has also resulted in lower prices. However, competition in the fixed market has tended to result in slightly higher average prices as tariffs are rebalanced. The net effect of higher penetration rates and lower prices has been an increase in the overall size of the sector. In countries that implemented comprehensive sector reforms early on, the sector-generated revenues were equivalent to about 6 percent of gross domestic product (GDP) on average. Countries that were late reformers, however, had sectors with total revenue equal to about 4.6 percent of GDP.

What lessons can be drawn from a decade of reform in the telecommunications sector in Africa? First, improvements in sector performance due to market liberalization take time to materialize and usually depend on the market's reaching a threshold level of competition. For example, increased competition among mobile operators significantly increased the availability of services and reduced prices in that segment of the market.

Yet countries that issued only two mobile licenses did not see as much benefit from competition as those that issued three or more. In general, the more mobile operators there are and the stiffer the competition for customers becomes, the faster the market will grow.

The second major lesson is that not all segments of the market are winners. In contrast to the major benefits that liberalization has conferred on the mobile segment of the market, competition in the fixed-line market has resulted in either static or declining numbers of fixed-line subscribers as customers switch from fixed-line to mobile services. The major exception is Nigeria, which aggressively liberalized its fixed-line market and has seen significant increases in the number of fixed-line subscribers. Nevertheless, the sector as a whole has grown dramatically as a result of liberalization. Customers have benefited from lower prices and better services while governments have benefited from higher tax revenues.

Financing

The rapid expansion of the telecommunications networks in Africa has required very high levels of investment. Between 1998 and 2008, an average of $5 billion a year was invested in Sub-Saharan Africa's telecommunications sector, amounting to about 1 percent of total GDP. The private sector accounted for most of this investment, which primarily targeted mobile infrastructure development following the liberalization of mobile markets.

This investment has not been distributed evenly across the continent. Nigeria and South Africa together account for more than 60 percent of the total network investment in Sub-Saharan Africa, with the remainder being distributed among the other countries in the region. The uneven distribution of investment corresponds to the size and relative wealth of these countries. In addition, in the case of some countries—such as Nigeria—it also reflects policy decisions made by the governments over the past decade. Countries that have promoted competition within the sector by encouraging new operators to enter the market have received higher levels of investment than countries that have limited competition.

What are the predominant sources of investment? The majority of sponsors of telecommunications investment in Sub-Saharan Africa originate in Africa itself, although much of the financing has been sourced from outside the continent. The only other region driving significant investment in the sector has been the Middle East. Historically, the telecommunications industry in Africa has seen relatively little involvement by investors from developed countries and from Asia. Recently, however, this has changed as Indian operators have begun to enter the market, and several recent privatizations were awarded to operators from developed countries.

Little of the capital invested in the sector has been in the form of equity from shareholders. The majority of investment is financed through debt, in the form of either bank loans or, to a lesser extent, issuance of bonds on local securities markets. More than half of the financing originates in Europe and North America, and 20 percent originates in the Middle East and North Africa. Yet the telecommunications sector in Africa has also successfully tapped local financial markets (to the extent that they exist) to fund investments. In some cases, telecommunications operators constitute a significant part of the total value of securities—both equity and debt—on exchanges. Local telecommunications borrowing has also been a big factor in the growth of loan syndication in African markets.

Despite the dominance of the private sector in telecommunications investment, in many African countries the public sector continues to play a role through its ownership of one of the operators. Approximately half of the countries retain full public ownership of one operator although several of these are considering privatization. At the same time, a few governments have begun reinvesting in the sector via national backbone projects.

In most cases, state-owned operators control only a small share of the market. Many of them have positive cash flows, so they do not place an undue cash burden on government budgets, but there is usually a significant opportunity cost in retaining them under state ownership. State ownership of a telecommunications operator has a long-term hidden cost, one that arises from biased regulatory and policy decisions designed to protect the government's investment. Governments would benefit by selling these operators both directly, through the revenues generated by the privatization, and indirectly, through the increased competition in the sector, resulting in higher tax revenues.

The public sector outside Africa is also a small but significant player in telecommunications investment in the region. For example, the government of Libya is directly investing in a number of operators in Africa through a regional investment company. In addition, development finance institutions have played a role in the development of the sector through financing many telecommunications investment projects. Traditional overseas development assistance, however, plays a very small role in the sector and is mainly focused on technical assistance rather than investment.

Future Investment Needs

Since the end of the 1990s, both the scale of investment in telecommunications and the speed of the networks' growth have been remarkable. Mobile networks cover 90 percent of Africa's urban population (94 percent if Ethiopia, the only large country in Africa that retains a state-owned monopoly, is excluded) and about half of the continent's rural population. Yet indications suggest that network coverage growth is slowing, and it is likely that some parts of the population live in areas in which mobile networks are not commercially viable. If this is the case, policy makers must ask themselves the following key questions: How far will the private sector drive network expansion? How much subsidy, if any, will be required to provide coverage to areas that would not otherwise be commercially viable?

The Cost of Providing Universal Voice Coverage

In the early stages of network growth in Africa, mobile operators concentrated on urban areas for two reasons. First, more high-income people live in cities and towns than in rural areas, so demand for services was higher there. Second, the high population density of urban areas allowed operators to achieve economies of scale in building networks, which translated into a lower average cost of providing mobile services. Nevertheless, the combination of license obligations and competitive pressures soon drove operators to expand their networks to cover small towns and rural areas.

Operators continue to compete with one another to extend their network coverage. It is unlikely, however, that network expansion will continue to cover 100 percent of Africa's population. In some rural areas, the population is so thinly distributed that it will be unprofitable to provide network coverage there for the foreseeable future. It is therefore important for policy makers to know how far network coverage will increase solely as a result of competition because this knowledge will guide future decisions on license obligations and universal access policies. By comparing the costs and the potential revenues of network coverage in each part of the region, one can develop an estimate of the future limit of network expansion. Expansion beyond this limit is likely to require some form of subsidy.

Recent developments in geographic information system technology have allowed a spatial analysis of the costs of network expansion and the associated potential revenues. By dividing Sub-Saharan Africa into small geographical units and comparing the cost of building mobile networks in these areas with their potential to generate revenue, one can develop a geographic view of where networks are commercially viable and where they would need to be subsidized.

This analysis estimates that 92 percent of the population of Sub-Saharan Africa is living in areas that are potentially commercially viable for mobile operators. The remaining 8 percent of the population lives in areas that are unprofitable. The estimated cost of providing coverage to these areas is just under $1 billion per year over nine years. For comparison, that total is about 20 percent of the total investment expenditure over the past 10 years and would result in universal network coverage in the continent.

Although the cost of providing universal network coverage does not appear to be excessive for Africa as a whole, this may not be the case for all countries. Countries in the region exhibit significant variation in the extent to which competition will drive network expansion and the level of subsidy that will be required to provide universal coverage. For example, the very large, thinly populated countries in Africa such as the Democratic Republic of Congo, Madagascar, and Zambia are likely to have much larger coverage gaps (that is, the proportion of the population living in unprofitable areas) than the smaller countries. The cost of providing universal coverage in these countries is therefore much greater.

The implications of this analysis are quite striking. The best way to achieve universal service is to encourage full liberalization and intensified competition, which will provide network coverage for more than 90 percent of Africa's population. Public subsidies or financial incentive schemes are therefore needed to cover only the most rural or difficult-to-reach areas of the continent. Providing such subsidies will not be cheap but, when compared with the total revenue generated by the sector, is feasible.

The Cost of Universal Broadband Internet Coverage

Broadband Internet in most African countries has been limited to major urban areas and to Internet cafés, businesses, and high-income residential customers. Network coverage is limited, prices are high, and connection speeds are lower than in other regions of the world. This situation is quickly changing, however, as operators and Internet service providers (ISPs) upgrade networks to provide wireless broadband services using 3G,

Effective market liberalization requires more than just issuing one or two mobile licenses. The liberalization process needs to go further and deeper. Governments should consider issuing more than two licenses covering all segments of the market. The evidence from Africa is clear that markets can sustain significant numbers of players in all segments. The licensing framework itself is also in need of reform. In particular, licensees should be free to invest and innovate in infrastructure and service delivery. The imperative for this is further driven by the process of technological convergence that is breaking down the traditional link between the network infrastructure and the service or services provided by it. The licensing framework needs to be reformed to match this evolution.

Even as markets have been liberalized, some governments have been exerting tighter controls over international gateway facilities. The temptation to do this should be avoided, because it is likely to decrease sector growth as well as government revenue generated by the sector.

The final part of the liberalization path is the privatization of the remaining state-owned enterprises. Although these operators are typically minor players in the sector, with small market shares, the presence of one state-owned operator generates a conflict of interest for the state, which has both a financial stake in its operator's success as well as regulatory responsibility for developing effective competition in the sector.

The second pillar of the sector reform program is the establishment of effective regulation. Regulators play a central role in ensuring that sector policy is implemented and that competition develops effectively. Many aspects are found in improving regulatory performance. Most important, regulators need to be institutionally independent of government. This is achieved by guaranteeing independent financing for the regulatory agency and having senior management appointed by the president, parliament, or a council of ministers rather than by sector management officials. In addition, regulators need to have sufficient legal powers to implement regulatory decisions without facing unending litigation by market players.

Regulators across the region face many of the same issues, from improving interconnection regulation to introducing measures that favor competition such as number portability and virtual mobile network operators. Facilities sharing is another common challenge faced by regulators and is seen as becoming increasingly important as networks are rolled out into rural areas and concerns are raised about the environmental impact of mobile tower infrastructure. Much of the drive

WiMAX, and other wireless broadband standards. It is also getting easier and cheaper for customers to access broadband Internet through handheld devices (such as personal digital assistants [PDAs]) and external wireless adapters for computers commonly referred to as "dongles." These technological innovations, together with the introduction of prepayment systems, mean that broadband is now beginning to grow in Africa.

Sector policy makers face two key questions: How far will this process go? Is wireless broadband commercially viable in Africa, and if so, will it extend beyond the major cities? To answer these questions, a modeling exercise was carried out. This is similar to that done for mobile voice networks: The costs of expanding networks and associated potential revenues were compared on a geographical basis to understand how far market forces alone would drive network expansion.

A baseline scenario was used in which wireless broadband networks offered a combination of personal and shared access to a relatively small subscriber base (1 percent broadband subscriber penetration in urban areas and 0.25 percent in rural areas), but with the expectation that far more people would be able to use the Internet at these shared facilities (such as Internet cafés) than actually subscribe for the service. This is the type of demand scenario that is currently being envisaged by many operators and ISPs in the region and is therefore likely to reflect the reality on the ground over the next few years. Nevertheless, many countries aspire to higher levels of Internet usage. An alternative scenario was therefore modeled in which target penetration rates are higher. Such levels of usage cannot be sustained on a commercial basis for the foreseeable future, however, because broadband networks cost too much to build compared with the revenues that they are likely to generate. This scenario analysis is used to estimate the level of subsidy that would be required to achieve these targets.

Under the first scenario—low penetration and shared access—broadband wireless networks would be commercially viable for about 75 percent of the population of Sub-Saharan Africa. This would be a major increase from current network coverage levels and consistent with forecasts from operators in the most advanced countries, such as Kenya and Nigeria. Expanding the networks to cover 100 percent of the population would require an additional $648 million per year over an eight-year period.

A more ambitious objective of mass-market personalized broadband access involves extending coverage into less commercially viable areas. This would require much greater levels of investment and would not be financially viable without extensive subsidies. With target broadband penetration rates of 20 percent in urban areas and 10 percent in rural areas, which is comparable to current OECD broadband penetration rates, the system would not be commercially viable even if subscribers were willing to pay $10 per month for the service. A total subsidy of about $10 billion per year would be required to make such a scenario commercially attractive to operators.

Broadband Internet is currently rare in Africa, but the advent of low-cost wireless broadband technologies—on both the network and the customer side—has brought wireless broadband within reach of much of Africa's population. This analysis indicates that in the short to medium term, the provision of wireless broadband services is commercially viable throughout much of the region. Still, at current costs and income levels in Africa, replicating the commercial success of broadband in high-income countries is unlikely to happen in the foreseeable future, unless governments are willing to provide financial support to the rollout of broadband networks and subsidize their use.

Policy Analysis and Conclusions

Full and effective market liberalization and sound regulation have been the major drivers of the rapid expansion of telecommunications networks and services in Africa. Yet the region as a whole still has a long way to go, and many countries lag far behind the region's best performers. The first key policy priority is therefore to complete the reform agenda. This will drive network expansion farther into rural areas and, at the same time, boost the development of more advanced segments of the market such as broadband Internet.

Some parts of Africa are always likely to remain commercially unviable because of difficult physical terrain or low revenue potential. In these areas, some form of direct incentive is likely to be required to achieve the region's policy objectives in the areas of basic voice and broadband Internet. The policy recommendations arising from this analysis can be divided into these two broad categories: (1) completing the reform agenda and (2) creating incentives for operators to meet evolving policy objectives.

Completing the Reform Agenda

Completing the reform agenda can be broken down into two parts: full liberalization and effective regulation. Both steps are aimed at promoting effective competition in the sector.

for this has originated in market forces with the establishment of tower companies to which mobile operators have been outsourcing their tower infrastructure and management, but a role also is seen for regulators in stimulating this process and ensuring that it does not adversely affect the development of competition.

Allocating bands of the radio spectrum is also an essential responsibility of the regulator, one that is particularly important in Africa because the wireline infrastructure is so limited. Regulators need to consider whether current global trends in radio-spectrum regulation—which are toward more flexibility and market-based mechanisms—could be profitably adopted in Africa.

Finally, an overarching issue facing all regulators in Africa is capacity building. The sector is changing rapidly, and the pressure on regulators is building as the sector grows. Regulators need to continually sustain their professional and institutional capacity so that they can carry out their mandate in this rapidly evolving environment. Regional regulatory institutions can support this process effectively through regional harmonization and capacity-building programs.

Creating Incentives for Operators to Meet Evolving Policy Objectives

Although liberalization and competition can help the sector meet many of its objectives, they are unlikely to bring about the fulfillment of *all* of them—including, for example, 100 percent mobile network coverage. In such cases, the public sector should provide incentives to companies to meet these objectives. These incentives can take many forms. At one end of the spectrum are changes to technical aspects of the regulatory framework; at the other end are direct financial subsidies. Some of the key incentives are discussed below.

Ensuring that affordable ICT services are provided in rural areas is a major policy priority. Regulators therefore need to consider carefully the many options for promoting this in the context of liberalized and increasingly competitive markets. Marginal areas of the country can be made more attractive to operators by reducing taxes on equipment and ICT services rather than, as has often been the case, elevating the level of taxes on the sector. Taxes can also be used to give direct incentives to provide coverage in rural areas.

Regulators can also encourage the use of low-cost technologies such as long-range base stations and solar power units for base stations to improve the financial viability of rural areas. Other cost-reduction strategies, such as facilities sharing, should be considered as well.

Regulators can also enhance the commercial viability of rural areas through revenue-enhancement strategies. For example, networks may be used to provide public services, with the government as an "anchor tenant." Networks can also be encouraged to offer value added services such as mobile banking that boost traffic and therefore revenues, making rural areas more commercially viable.

Universal service funds are the traditional means of boosting network coverage or giving incentives to operators to provide services in areas that would be otherwise unprofitable. These funds are financed through levies on the sector and then ideally allocated through some form of minimum-subsidy competition. Despite sounding good in theory, however, such schemes have often been less than satisfactory in practice, and, in many cases, funds collected for universal services remain undisbursed. For assurance that these funds have the maximum impact, they should be limited in size and should be focused only on areas of the country that are not otherwise commercially viable, ideally when network coverage growth is starting to slow. They should also be implemented only in fully liberalized markets in which there is effective competition and be designed to promote competition rather than displace it.

Broadband is a new area for the telecommunications sector for Africa, so the strategy for promoting its development is less clear-cut. Because it is a complex product, policy needs to cover all parts of the value chain to be successful. The development of many new, for the most part privately financed, submarine fiber-optic cables has resulted in large increases in international bandwidth available to Africa. This, together with diversity of the ownership, is likely to result in greater competition and lower prices. In addition, international competition is growing in the manufacture of wireless broadband access network equipment. Here prices are already falling quickly, a trend that is likely to continue as operators in high-income countries step up their investments in wireless broadband networks. Few financial barriers exist to entry into this segment of the market, so governments should facilitate market entry and competition by removing regulatory barriers such as license restrictions and spectrum constraints.

Backbone networks are essential for the provision of broadband Internet, but they are currently undeveloped in many African countries, partly as a result of the high investment costs and excess of regulatory constraints. Removing these constraints should encourage investment in the completion of these networks, particularly the trunk routes between major urban areas and submarine-cable-landing stations. Beyond these areas, some form of public financial incentive is likely to be needed to infrastructure remain very low. Even in countries where the average telecommunications penetration rate is high, major disparities are often seen between rural and urban areas. Rural coverage levels are growing steadily, but they still lag far behind availability in cities and towns. Nearly all urban areas in Africa are now covered by mobile networks, but less than half of the rural population lives within reach of a network.

The ICT revolution has, to date, been largely one of basic voice communications. Access to the Internet is still very limited. Today, there are about 19 million broadband Internet subscribers in the entire Sub-Saharan African region,[2] about 6 percent of the total number of telephone subscribers. However, these are concentrated in just two African countries—Nigeria and South Africa—which together account for 15.7 million broadband subscribers, more than 80 percent of the Sub-Saharan African total.

Prices for ICT services in Sub-Saharan Africa are generally high relative to other regions, but, in most countries, they have been falling as markets have reformed and competition has developed. Some prices have fallen faster than others. The average price of an international call from Sub-Saharan Africa to the United States, for example, fell by 57 percent between 2000 and 2008. Mobile call prices have fallen steadily across the continent, dramatically in some cases. Internet access, however, remains prohibitively expensive in most countries: Prices in Sub-Saharan Africa are much higher than in high-income countries in other regions of the world or even the middle-income countries of North Africa. Not only are Internet prices very high, but quality is generally very poor.

Africa's telecommunications infrastructure has grown rapidly since the end of the 1990s. Most of this growth has been in mobile networks, which have used predominantly wireless technologies. More recently, fiber-optic networks have been developed to handle the increased traffic levels generated by a large number of voice subscribers and an increasing number of broadband Internet users. This terrestrial infrastructure is also being complemented by the offshore submarine fiber-optic cables that are increasingly connecting Africa to the rest of the world. This ongoing growth and transformation of Africa's communications infrastructure is driving change in the retail market, as operators are able to provide new services and lower prices.

Access: Burgeoning, at Least for Mobile Telephony

As previously stated, access to telecommunications services has increased dramatically. Between 1998 and 2008, 247 million mobile subscribers joined the networks, bringing the total number of telephone subscribers (fixed and mobile) to 263 million. This increase in the number of subscribers raised mobile penetration rates from 0.6 percent to 32 percent (figure 2.1).[3] The market has also become more evenly distributed across the region. For example, South Africa accounted for 86 percent of subscribers in 1998, but by 2008, that figure dropped to 18 percent. In 2008, Nigeria overtook South Africa as the biggest mobile market in Sub-Saharan Africa, with one-quarter of the region's total number of mobile subscribers.

Other telecommunications services are much less widely available in Africa. Broadband Internet penetration rates increased from zero at the beginning of the decade to about 19 million in 2010. Relative to the population as a whole, this is still very small, at about 2 percent,[4] but it is growing quickly, increasing at an average rate of 200 percent per year between 2005 and 2009.

By contrast, the fixed-line segment of the market in Sub-Saharan Africa remains small, having barely increased in recent years. The fixed-line penetration rate was 1.3 subscribers per 100 people in 1998; it rose to 1.5 in 2007, then fell to 1.4 in 2008.

The ICT market in Africa has seen rapid growth across the continent since the late 1990s, but at the regional level, the market is dominated by

Figure 2.1 Telecommunications Subscribers, Sub-Saharan Africa, 1998–2008

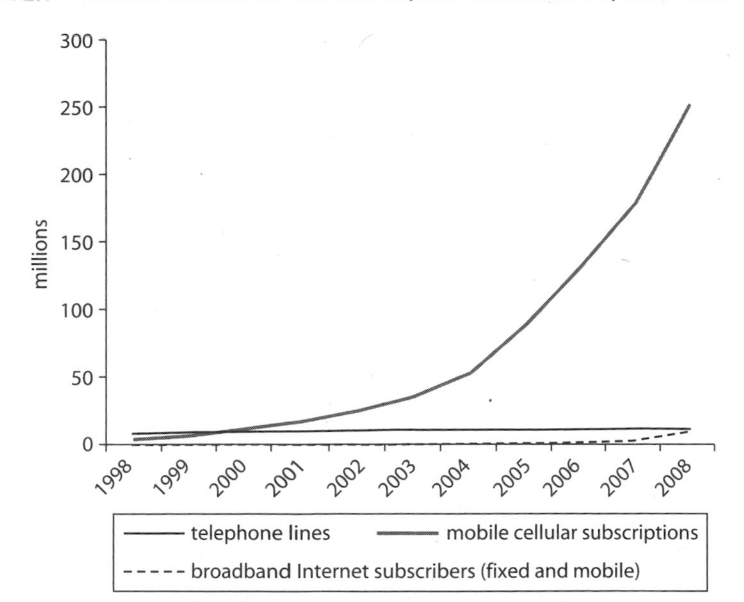

Sources: ITU 2010; Wireless Intelligence; World Bank Development Data Platform.

Figure 2.4 Average Penetration Rates, Top 10 Countries in Sub-Saharan Africa versus Bottom 10, 2008

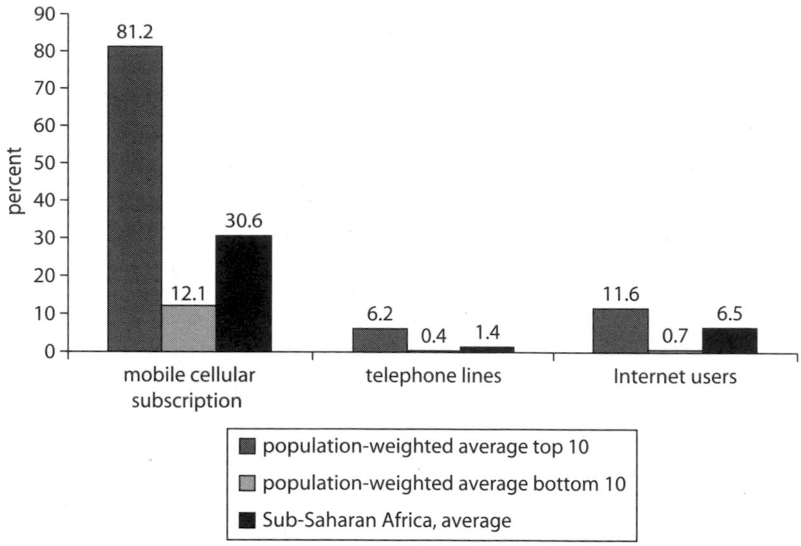

Sources: ITU 2010; World Bank Development Data Platform.

Major disparities in access to ICT services are also found within countries—between rich and poor, rural and urban. Less than 5 percent of rural African households have access to a fixed-line telephone, compared with about 17 percent of urban residents (figure 2.5, panel a). Similarly, disparities in access to mobile services are enormous: Close to 30 percent of all cellular subscriptions belong to households in the wealthiest quintile, compared with less than 2 percent for the poorest quintile (figure 2.5, panel b).

In the past, public phones have been an important alternative form of access to ICT services. But policies that promote access to pay phones have been complemented and, in many cases, replaced by the resale of mobile phone services, either through formal operator schemes such as Village Phone or informally through individual resale. As the data for Kenya show, the numbers of mobile pay phones overtook those of fixed pay phones in the mid-2000s but have recently begun to decline as individual access to mobile phones has increased (figure 2.6).

Around the world, the major new development in the ICT sector has been the growth of broadband Internet. Seven out of every 100 people in

Figure 2.5 Access to Telephones by Rural/Urban and Income Quintile, 2007

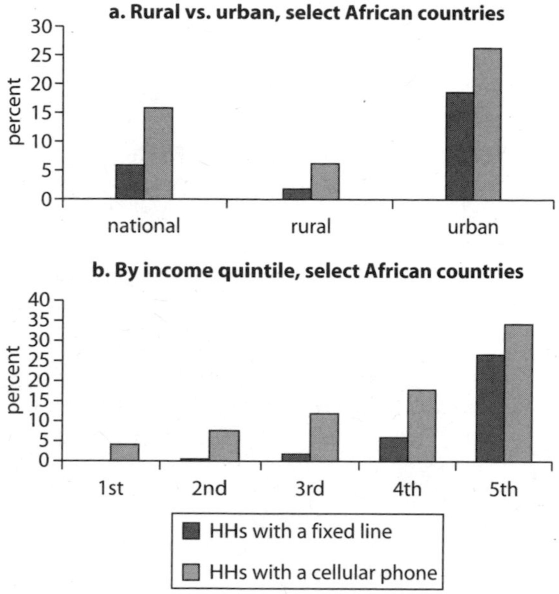

a. Rural vs. urban, select African countries

national rural urban

b. By income quintile, select African countries

1st 2nd 3rd 4th 5th

HHs with a fixed line

HHs with a cellular phone

Source: Ampah and others 2009.

Note: Simple average of countries with data. HH = households. The sample of countries for which data are available is Benin, Burkina Faso, Cameroon, Cape Verde, Central African Republic, Chad, the Comoros, the Democratic Republic of Congo, the Republic of Congo, Côte d'Ivoire, Ethiopia, Gabon, Ghana, Guinea, Kenya, Lesotho, Madagascar, Malawi, Mali, Mauritania, Mozambique, Namibia, Niger, Nigeria, Rwanda, Senegal, South Africa, Sudan, Tanzania, Togo, Uganda, Zambia, and Zimbabwe. Not all types of data are available for every country.

the world used the Internet in 2000 (World Bank 2010). It is estimated that now 27 percent of the world's population use the Internet,[6] and in high-income countries, usage rates are even higher. The number of broadband subscriptions in the OECD countries, for example, stood at 22 per 100 people in June 2009 (OECD 2009a). Many Internet connections are shared—for example, by families—and so the number of individual Internet users should be significantly higher.

Usage patterns have also changed. Once considered a luxury only available to the small proportion of the population that could afford it, the Internet has become increasingly integrated into societies and economies around the world. Developing countries are now seeing the economic benefits of the Internet through improved productivity, expanding economic development opportunities, and better service delivery (Qiang and Rossotto 2009).

Figure 2.6 Public Pay Phones in Kenya, 2001/02–2007/08

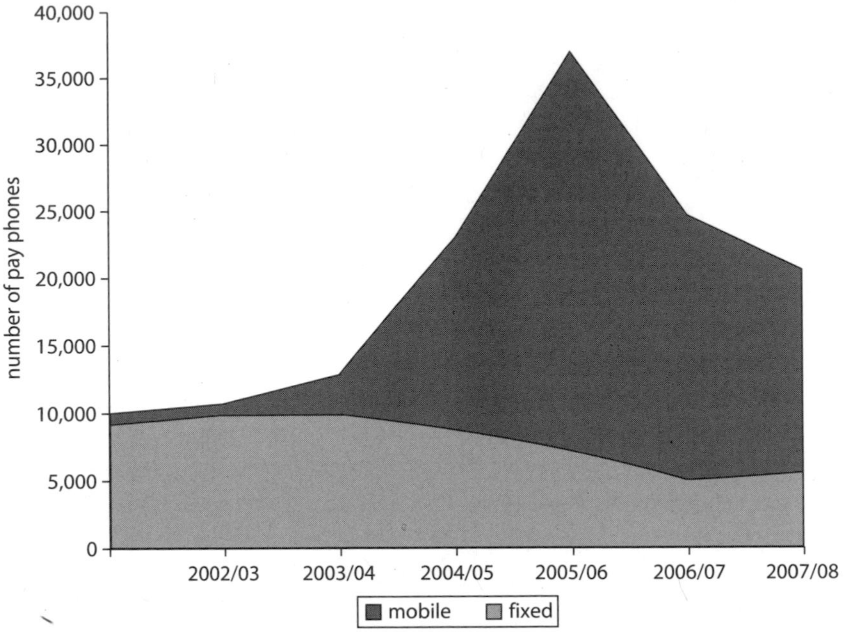

Source: Adapted from the Communications Commission of Kenya 2009.

Internet usage in Sub-Saharan Africa has been slow to take off and lags far behind the rest of the world, but this is starting to change with the emergence of global standards for wireless broadband networks, increases in international bandwidth available to Africa, and reductions in the price of network equipment and end-user devices. By the beginning of 2010, broadband Internet was available in most countries in Sub-Saharan Africa, and the total number of subscribers rose from 0 to more than 19 million between 2000 and the beginning of 2010. The current subscriber penetration rate is about 2 percent (World Bank 2010; Wireless Intelligence).[7] The vast majority of subscribers—over 95 percent—access broadband via wireless technologies; there are fewer than a million fixed-wireline broadband Internet subscribers (World Bank 2010). This picture is somewhat misleading, however, because the subscriber base is concentrated in a few countries: 15.7 million subscribers (81 percent of Sub-Saharan Africa's total broadband subscriber base) are in Nigeria and South Africa alone (figure 2.7).

Figure 2.7 Wireless Broadband Subscriber Growth (2000–10), Top 10 Countries in Sub-Saharan Africa

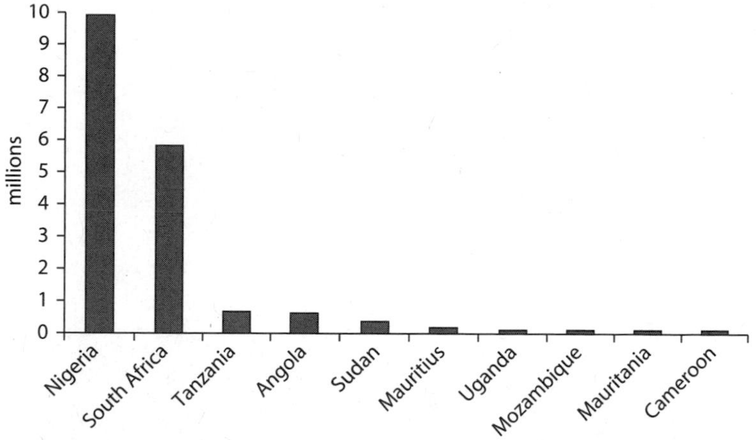

Sources: Wireless Intelligence; World Bank staff analysis.

Broadband usage trends in Nigeria are worth noting (box 2.1). There the Internet is primarily accessed through wireless technologies. Voice subscribers have upgraded from voice-only to voice-and-broadband services in large numbers, and today, 11 percent of mobile subscribers are able to access broadband Internet through their handset. Other people are using fixed Code Division Multiple Access (CDMA) and Worldwide Interoperability for Microwave Access (WiMAX) technologies for broadband access. The success of wireless broadband Internet is already being seen elsewhere in Africa. In Kenya, for example, mobile broadband started in 2008 with the launch of Safaricom third generation (3G) services. By 2010, Safaricom reported that its 3G subscriber base had risen to 3.4 million or 16 percent of the total number of mobile subscribers in Kenya (Business Monitor International 2010b). These examples perhaps point to the direction that other countries in Africa will move as the capacity of networks improves, broadband services are launched, and the cost of broadband handsets falls.

Prices: Falling, Where Competition Is the Rule

The price of a telecommunications service is a key indicator of its accessibility, in particular for the poor.[8] The price of fixed-line telecommunications services varies widely across countries. Monthly baskets for a

Box 2.1

The Challenge of Estimating the Number of Broadband Users in Africa: The Case of Nigeria

Estimates of the number of broadband subscribers in Africa vary widely. Nigeria provides a good example of the uncertainties inherent in such estimates. Wireless Intelligence estimates that there were 8.4 million CDMA evolution–data optimized subscribers in Nigeria in 2009 and a further 350,000 3G subscribers, yielding a total broadband subscriber base of 8.8 million in 2009 and 9.9 million in 2010. That estimate appears to be broadly consistent with survey data from Nigeria.

Citing Nielsen Online, Lange (2010, p. 81) writes that "some 7.3 million people accessed the Internet via their mobile phones during the second and third quarters of 2008, an increase of 25 percent."

Business Monitor International, on the one hand, gives a much lower estimate of the broadband market in Nigeria, judging that there were 1.5 million 3G subscribers in Nigeria at the end of 2009 but only 3.4 million broadband subscribers overall (Business Monitor International 2010a).

Although estimates of subscribers in Nigeria range widely, there is broad agreement on the number of users, which most estimates put at about 30 percent of the population, the second highest rate in Africa after Mauritius.

The difficulty of estimating the number of broadband subscribers arises from several sources. The first is basic data availability. Data on the number of broadband subscribers are commercially sensitive; therefore, operators typically do not make the information public unless they are required to by the regulator. Disclosure is a common requirement for basic services such as voice telephony, but because broadband is relatively new in Africa, regulators have generally not yet introduced requirements to report broadband subscriber numbers in a systematic way. A second source of difficulty springs from the definitions used. The concept of a "broadband subscriber" originates in fixed-line broadband connections, which typically are provided under a contract lasting a year or more and entailing payment of a regular bill, irrespective of how much the customer uses the service. In such cases, a broadband subscriber is a well-defined concept and easy to measure.

Mobile broadband in Africa, on the other hand, is increasingly a prepaid service, as is the case for mobile voice services. A customer may have a broadband-enabled handset or computer interface but may not use it for broadband services.

(continued next page)

Box 2.1 *(continued)*

Industry-wide standards for definitions of broadband subscribers in a predominantly prepaid market have not yet been developed in a systematic way. Therefore, considerable uncertainty exists about the true number of customers who are active users of broadband services.

Despite this uncertainty about the total number of broadband users, it is clear that the number of Internet users and subscribers is growing rapidly as prices for services and devices fall and competition develops at all levels of the market.

Source: Authors.

subscription-based, conventional fixed-line telephone service averaged $11 in 2009, ranging from less than $1 in Ethiopia to $24 in Senegal and Zambia (figure 2.8). Countries such as Ethiopia that have monopoly operators tend to have relatively low average prices. In countries that have privatized their fixed-line operator (for example, Côte d'Ivoire, Senegal, and South Africa), fixed-line prices have risen. In Nigeria, which has the most competitive fixed-line market in Sub-Saharan Africa, the fixed-line price basket is low—about $6 per month.

The structure of fixed-line charges also varies by country. Fixed-line services usually have a two-part price structure that includes a fixed monthly subscription charge and a usage-based charge. The share of subscription charges in the total monthly price varies significantly. The high monthly fixed cost of subscribing to a wired telephone network limits access to ICT services, particularly for low-income households.

The structure of charges for mobile services differs from that of fixed-line services. Ninety-seven percent of mobile subscribers in Sub-Saharan Africa have a prepaid package with no fixed monthly subscription charge. The average monthly prepaid mobile basket in 2009 was $10 but ranged from $2 to $15 (figure 2.9). Note that the average price of mobile services is not significantly different from that of fixed-line services. This indicates that the limited usage of fixed-line services is not a result of their cost, but rather other factors, such as the availability of services and the charging structure (that is, usually postpaid).

The price of mobile services has dropped as networks have expanded, competition has intensified, and operators have reduced their prices to pursue new customers from low-income households.

Broadly, mobile prices over time are assessed in one of two ways: (1) the price basket methodology, which considers the total cost of a bundle of different types of calls, and (2) a calculation of average total revenue per minute. Trends in both of these measures for Africa are shown in figure 2.10.

As prices have fallen and networks have expanded into rural areas, more marginal subscribers—who tend to make fewer calls than existing customers—have joined the networks, further pushing down average revenues. Together, these effects have reduced the average revenue generated per subscriber. The mobile monthly average revenue per user (ARPU) in Sub-Saharan Africa stood at $8 in 2010, less than one-fifth of the $42 figure in 2000 (figure 2.11). Median mobile ARPU in Sub-Saharan Africa and North Africa is similar.

Despite the clear downward trend in mobile prices in Sub-Saharan Africa, average prices have been higher overall than in developing countries in other parts of the world. One reason for this is the market structure and the extent of competition (discussed in detail in chapter 3), but other factors also contribute to higher mobile prices, including interconnection charges, taxes, and high costs of inputs such as energy.

Mobile operators are subject to various taxes that feed through into higher prices. Countries often levy import duties on network equipment and mobile handsets. Value added tax (VAT) or sales tax, ranging between 5 and 23 percent, are usually levied on the sale of prepaid credit. In addition, some countries, particularly in East Africa, charge an excise tax on calls (figure 2.12). Together, import duties, VATs/sales taxes, and excise taxes can significantly increase the cost of mobile ownership.

Energy shortages also affect prices. Network operators must operate their own sources of power for mobile base stations and other telecommunications equipment. Self-generated power costs between $0.18 and $0.48 per kilowatt-hour (kWh), compared with $0.10 per kWh (Foster and Steinbuks 2009)[9] for grid power. These higher costs raise the prices paid by customers.[10]

The price of international calls in Sub-Saharan Africa also remains high. There was a dramatic drop between 2000 and 2006, when the price of a call to the United States fell by more than half (figure 2.13), but since 2006, prices have stagnated.

The price of international calls varies widely across countries (figure 2.14), to a large extent because of the degree of competition in this segment of the market. This varies across countries and particularly

Figure 2.10 Trends in Mobile Prices in Africa

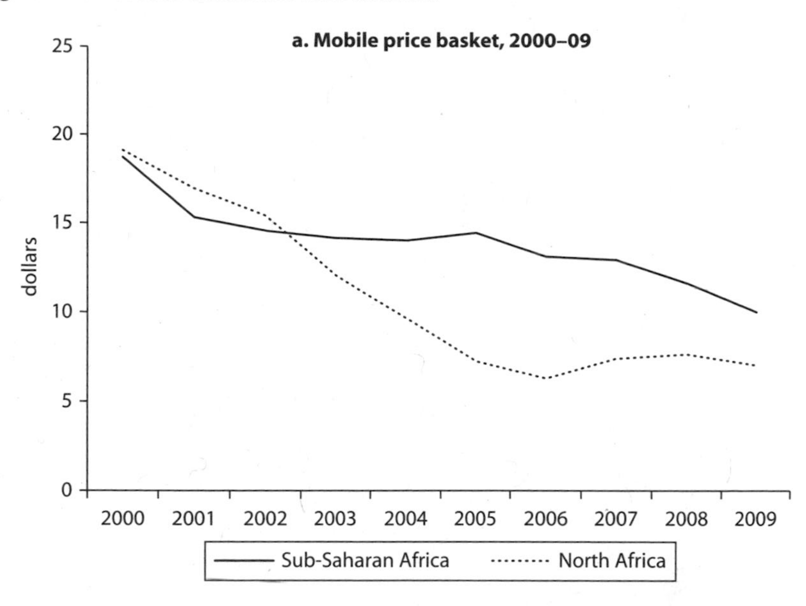

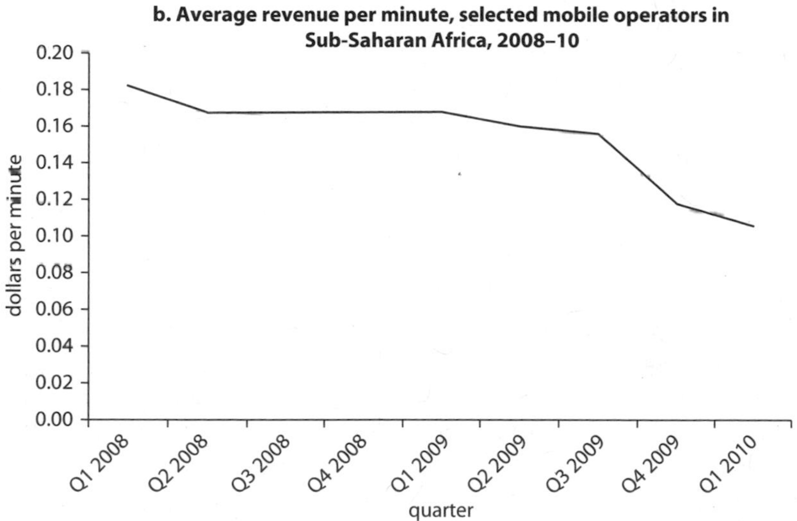

Sources: (a) Ampah and others 2009, updated. (b) Wireless Intelligence.

Notes: In panel a, the mobile cellular subbasket is based on the 2001 methodology of the OECD "low-user basket" (OECD 2002). It represents the price of a standard basket of mobile monthly usage in dollars determined by the OECD for 25 outgoing calls per month (on-net, off-net, and to a fixed line; and for peak, off-peak, and weekend periods, according to predetermined ratios) plus 30 text messages. Since the price of calls often depends on the time of day or week they are made, peak, off-peak, and weekend periods are taken into consideration. In panel b, only limited data on mobile operators' revenue per minute in Africa are publicly available. These data are from operators in Angola, Cape Verde, Ghana, Kenya, Lesotho, Mozambique, Nigeria, South Africa, and Tanzania. Q = quarter.

Figure 2.11 Median Mobile Monthly Average Revenue per User, 2000–10

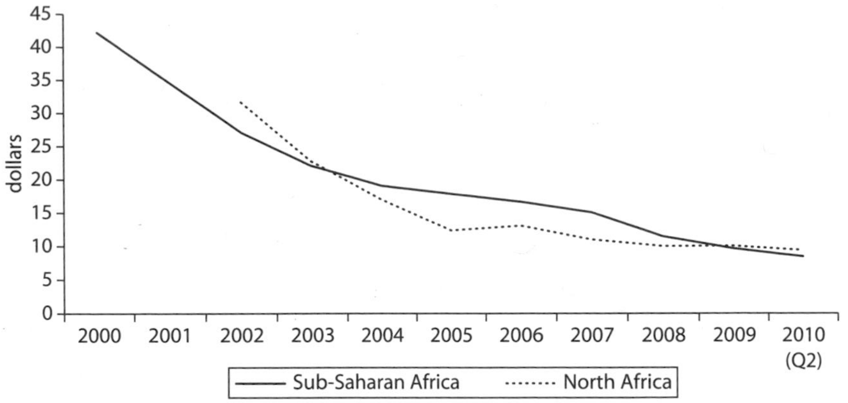

Source: Wireless Intelligence.
Note: Q2 = second quarter.

affects the price of international calls. Before the introduction of competition, many operators had tariff structures that were unbalanced (that is, local tariffs were set below cost while international calls were set above cost). Competition has forced operators to rebalance tariffs by bringing them more in line with costs. This usually involves reducing the price of international calls and increasing the cost of monthly subscriptions and local calls.

Two main sources of competition exist in the market for international calls. If mobile operators are allowed to build their own international gateway facilities to carry international traffic, competition develops and prices fall. Where mobile operators are required to use the incumbent operator to carry international traffic, competition is less intense and prices remain high. International call prices are also particularly affected by the activities of Voice over Internet Protocol (VoIP) operators, who often target the international call segment of the market.

The prices of international calls within Africa are often higher than the prices of calls from Africa to the United States. The median price of international calls within Sub-Saharan Africa was one-third higher than international calls to the United States, but, again, significant variation is seen across countries (figure 2.15).

Figure 2.12 Value Added and Excise Taxes on Mobile Communications Services, 2007

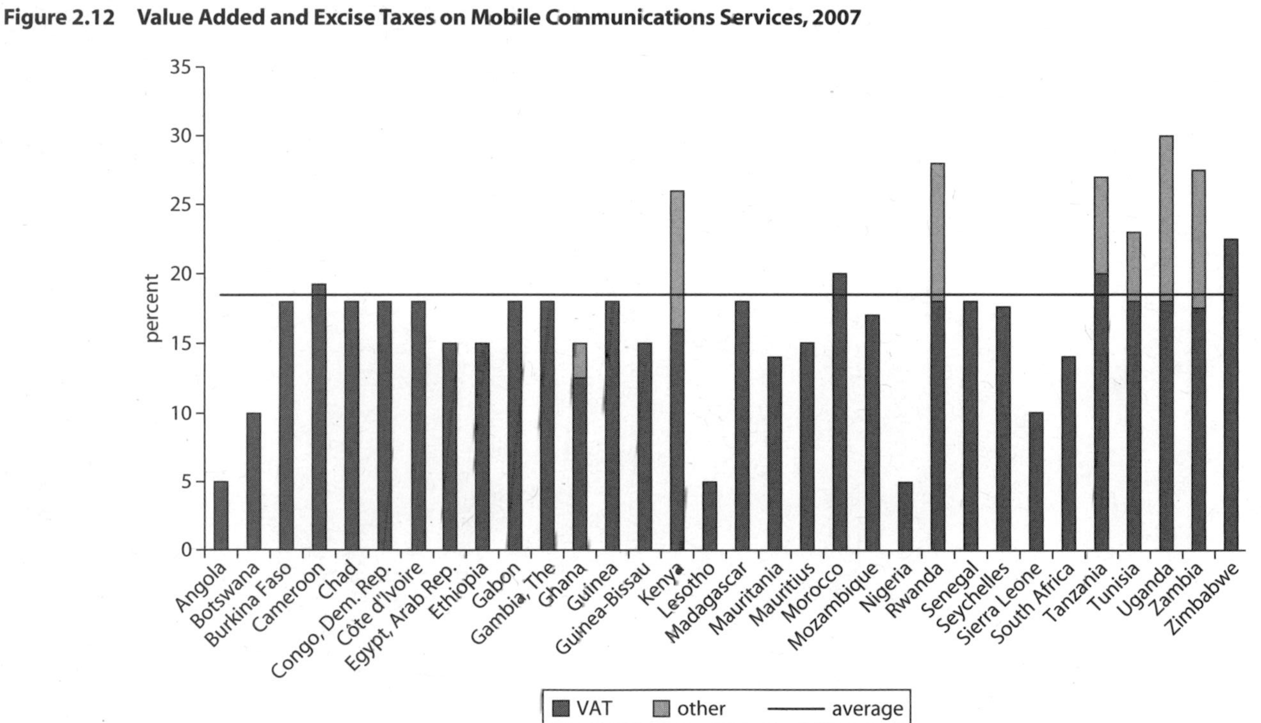

Source: Adapted from GSMA 2007.
Note: VAT = value added tax.

Figure 2.13 Average (Mean) Price of a One-Minute Peak-Rate Call to the United States from Countries in Sub-Saharan Africa, 2000–08

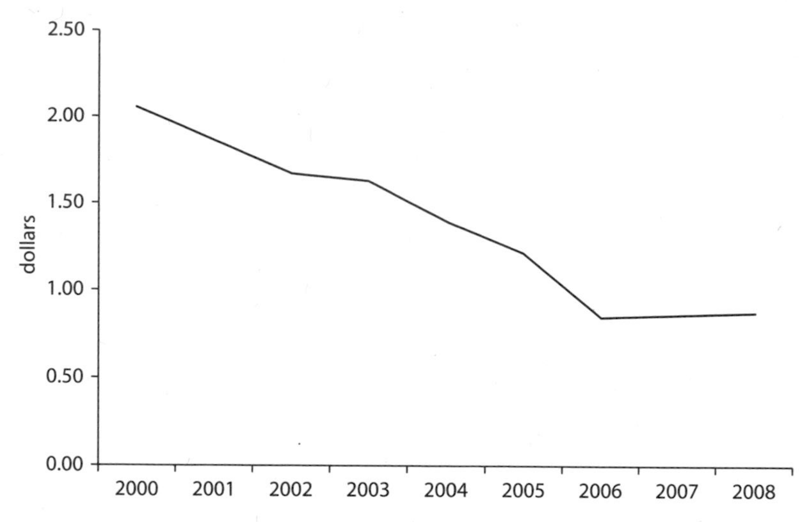

Source: Ampah and others 2009, updated.
Note: Peak rate, including taxes. The country sample includes Djibouti.

Another feature of intra-African call prices is that they tend to be lower within trade blocs than between them, although ECOWAS (the Economic Community of West African States) is an exception (figure 2.16).

The Internet is very expensive in Africa. In 2010, the average price of fixed broadband in Sub-Saharan Africa was $139 per month—high by international standards but significantly lower than in 2008, when the average price was over $300 per month. This average is skewed by countries in which the monthly cost of broadband is over $500. The median price in 2010 was $56 (figure 2.17).

Fixed broadband is much more expensive in Africa than in other parts of the world, both in absolute terms and relative to average incomes, but this gap has started to close as broadband prices in Africa fall (figure 2.18).

The price of broadband Internet provided via wireless networks is lower than over fixed lines but is still high by international standards. The median price of mobile broadband (3G) in Sub-Saharan Africa is $41 per month, which is less than the price of fixed broadband but is nearly four times the price of mobile broadband in North Africa. A dramatic

Figure 2.14 Price of One-Minute Peak-Rate Call to the United States from Countries in Sub-Saharan Africa, 2010

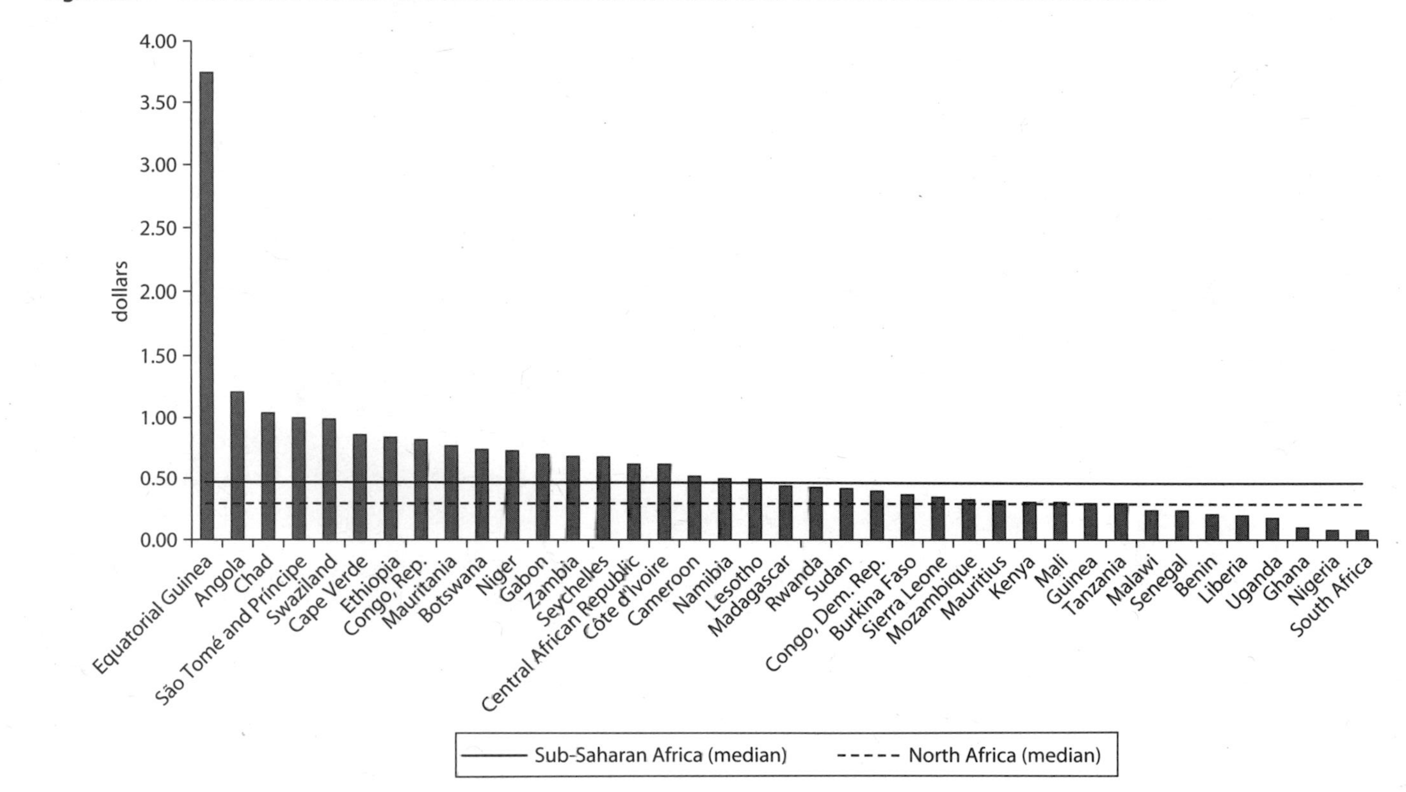

Source: Data from operator websites.
Note: Peak rate, including taxes. Calculated using annual average nominal exchange rate.

Figure 2.15 Price of One-Minute Peak-Rate International Call within Africa, 2010

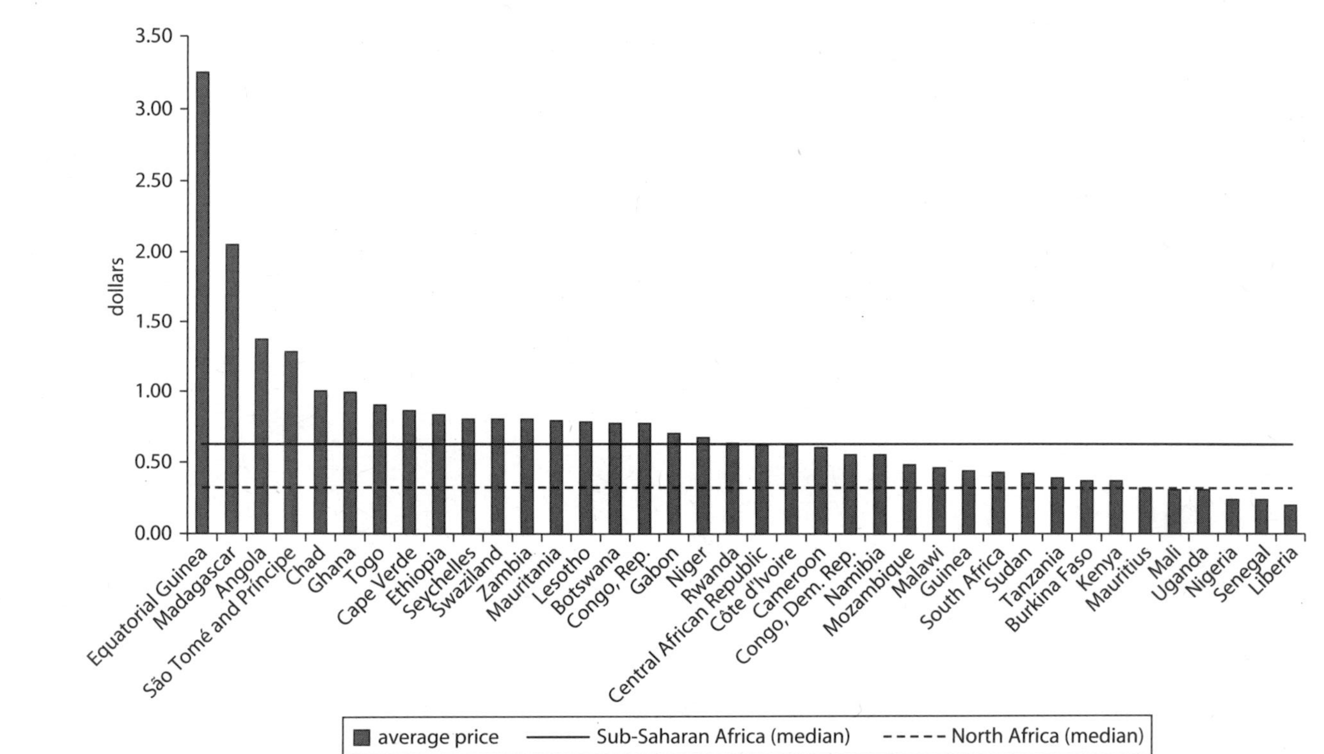

Source: Data from operator websites.
Note: Price for each country computed as the average of the price of a peak-rate call to all other African countries.

Figure 2.16 Price of One-Minute Peak-Rate Call within and outside Sub-Saharan African Trade Agreements, 2010

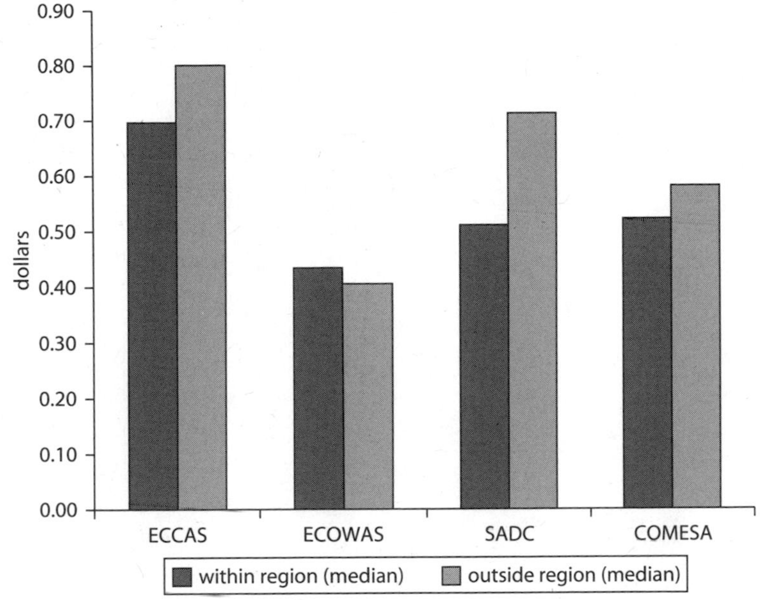

Source: Data from operator websites.
Note: The country with the most expensive tariff was excluded from the calculation to avoid distortion due to outliers. COMESA = Common Market for Eastern and Southern Africa; ECCAS = Economic Community of Central African States; ECOWAS = Economic Community of West African States; SADC = Southern African Development Community.

difference is also found in prices between countries: The most expensive mobile broadband in Sub-Saharan Africa was 47 times the price of the cheapest (figure 2.19).

Quality: Reliability Is a Problem

After physical availability and price, quality is an important issue determining whether people can benefit from ICT services. The quality of fixed-line services in Africa is poor. The standard measure of service quality for these services is the number of faults per 100 main lines per year. Most Sub-Saharan countries lack recent data, but for countries that report this figure, the average value was 69 in 2005. In other words, almost 7 out of 10 fixed lines were out of service at some point during the year. By comparison, the average figure for 14 OECD countries in

Figure 2.17 Fixed Broadband Price, 2010

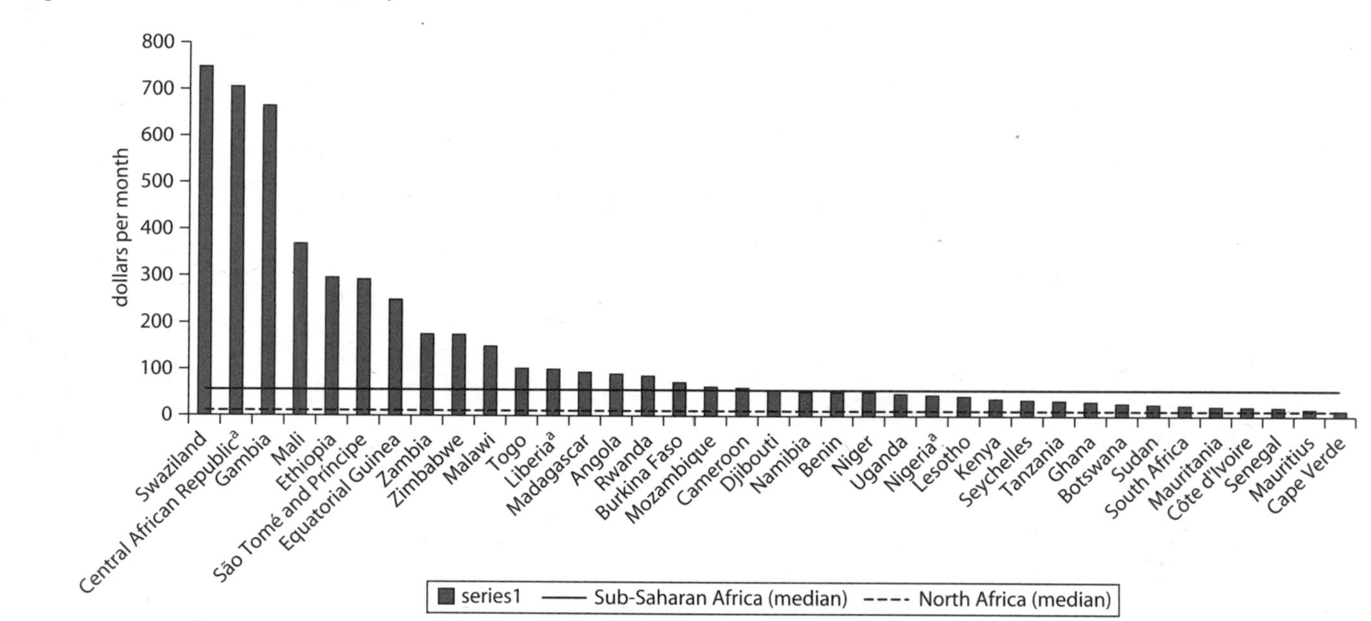

Source: Adapted from operator websites.

Note: Packages providing at least 256 kilobits per second and gigabits capacity at 20 hours per month. Prices valid August 1, 2010. Converted to dollars using 2009 annual average exchange rate.

a. Fixed wireless.

Figure 2.18 Fixed Broadband Internet Subbasket by Region and Level of Development, 2008

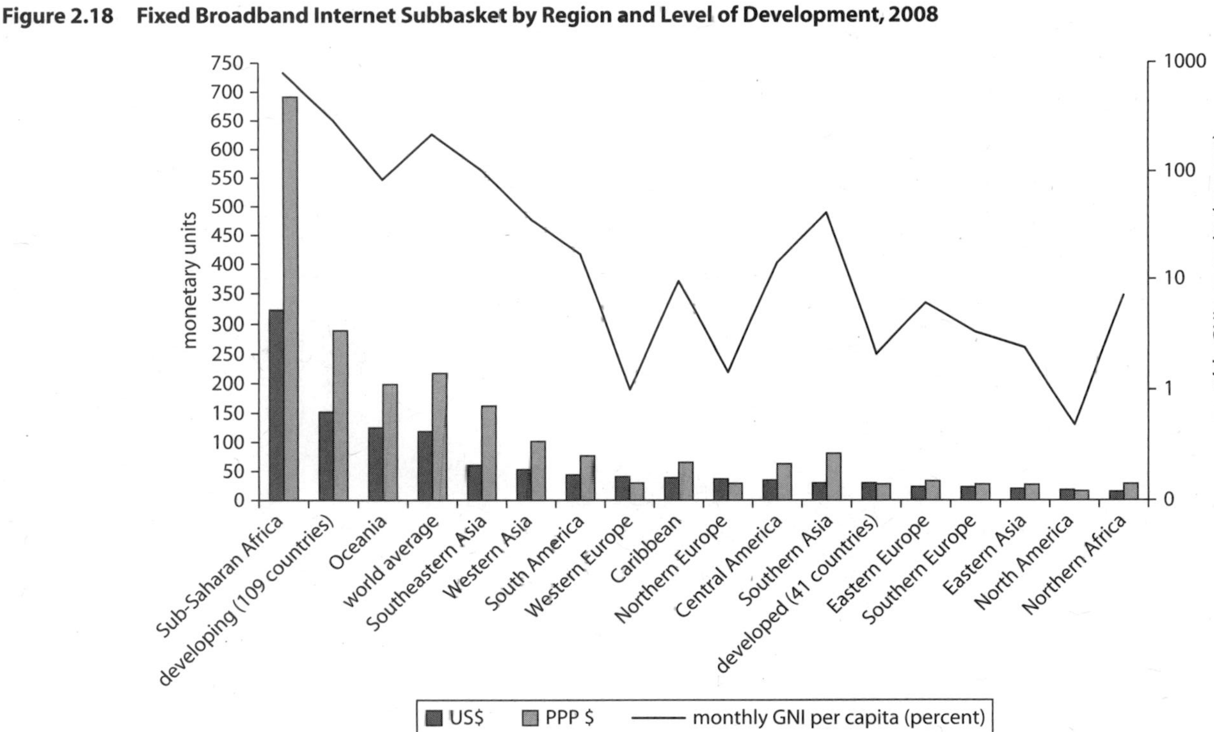

Source: ITU 2010.
Note: Logarithmic scale on right-hand ***y*** axis. GNI = gross national income; PPP = purchasing power parity.

Figure 2.19 Monthly Mobile Broadband Prices, 2010

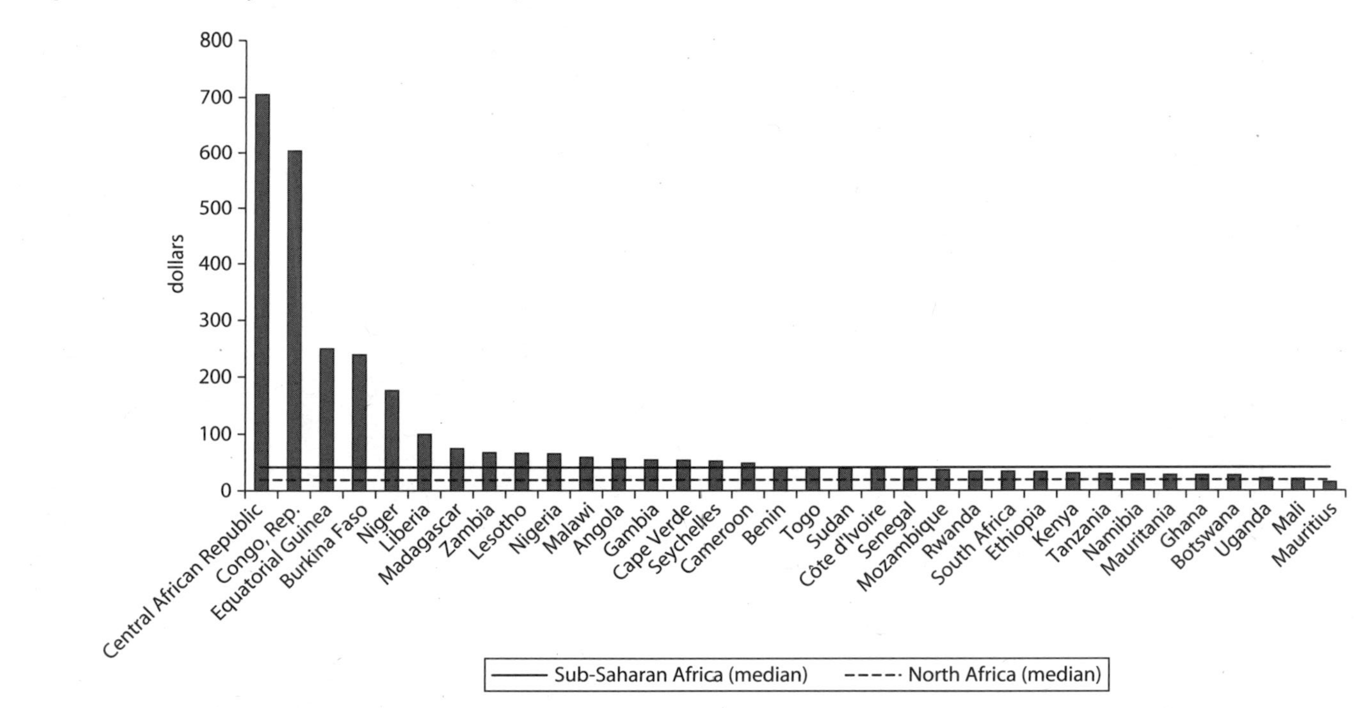

Source: Data from operator websites.
Note: Monthly subscription for 1 gigabit per month (download) of usage.

2003 was only 1 in 10. Canada and the Republic of Korea reported fault rates of 1 in 100 (OECD 2005).

The quality of the services provided by mobile networks is becoming a major issue in many African countries, because customers often face difficulty in connecting calls or find their calls cut off prematurely (Adegoke, Babalola, and Balogun 2008). Regulatory authorities in Sub-Saharan Africa do not systematically publish quality-of-service statistics, so quantitative comparisons are difficult. Senegal's Telecommunications and Post Regulatory Authority carried out a quality survey in October–November 2006 across four applications: voice, short message service (SMS, or texting), data (general packet radio service), and interoperator calls (ARTP 2007). For voice, the survey assessed audio quality and the ease of establishing and maintaining a call for two minutes. For SMS, the length of time to receive the message was tested. The call-failure rate between networks was measured for interoperator calls. The success rate of transmitting a data message was also measured. The results were aggregated into a single indicator ranging from 0 percent to 100 percent; higher values indicated better quality. The overall synthetic indicator for voice quality for Senegal's two mobile networks ranged from acceptable to perfect, with ratings between 80 percent and 94 percent. Meanwhile, the Uganda Communications Commission (UCC) began collecting quality-of-service statistics from 2007. In 2008, it reported a call-blocking rate of 1.8 percent (UCC 2008).[11] This might not seem high, but note that it is an average, whereas call blocking typically occurs at peak times. At such times, networks become congested and customers are likely to experience much higher rates of call blocking.

Several measures of Internet service quality are in use. One is the speed at which customers can download and upload data. There has been a dramatic increase in Internet speeds at the level of the access network. In the early days of the Internet, when dial-up was the predominant form of access, typical speeds were about 56 kilobits per second (Kbps). The introduction of broadband has revolutionized this, and connection speeds have increased rapidly. The definition of broadband used by regulators and governments has therefore also evolved. The OECD requires speeds of at least 256 Kbps for a connection to be classified as broadband (OECD 2008). In practice, typical broadband access speeds in advanced telecommunications markets are well above this. The average broadband speed in the OECD countries in 2008 was about 17 megabits per second (Mbps) (OECD 2009b), and access speeds of up to 100 Mbps are now common. The U.S. Federal Communications Commission (FCC) once defined broadband as an Internet service capable of providing speeds in excess of 200 Kbps. This definition was changed in 2010 to a download speed of 4 Mbps and an upload speed of 1 Mbps (FCC 2010).

In Sub-Saharan Africa, Internet access speeds are much lower. A key driver of this has been the availability of international Internet connectivity. As a whole, the region had approximately 25 gigabits per second (Gbps) of international bandwidth in 2008, of which just less than half was in two countries: Senegal and South Africa. The amount of bandwidth available per person varies dramatically across countries, as does the quality of Internet available to customers (figure 2.20; note the logarithmic scale).

It is important to put the amount of bandwidth available in Africa (shown in figure 2.20) into a global perspective. Between 2002 and 2009, the average per capita international bandwidth capacity in high-income countries rose from 2.6 Mbps per capita (Mbps/cap) to 50.4 Mbps/cap. In Latin America and the Caribbean, that figure rose from 0.09 to 4.90 Mbps/cap over the same period, but in Sub-Saharan Africa, by 2009, international bandwidth had reached only 0.06 Mbps/cap (TeleGeography 2010). More bandwidth is available in Egypt than in the whole of Sub-Saharan Africa, and the per capita bandwidth of South Asia and Latin America and the Caribbean is 7 and 83 times that of Sub-Saharan Africa, respectively. This chronic lack of international bandwidth available in Africa is one of the key reasons for the very low quality of broadband Internet experienced by most users.

Infrastructure: Bottlenecks Impede Growth

The telecommunications network infrastructure in Africa, as in other parts of the world, has evolved as technology has changed and customers have demanded new services. The type of infrastructure that operators install is a major determinant of the types of services that are available to customers, the quality of those services, and the dynamics of the competition among operators.

As noted in chapter 1, telecommunications infrastructure consists of interconnected networks that carry different types of traffic. The familiar division into "fixed" and "mobile" networks, which reflects the divide between fixed-line networks and the wireless networks that began to emerge in the 1980s, really applies only to the "last mile"—that is, the link between the network and the customer, because much of a mobile network actually consists of fixed (usually fiber-based) links. Nor is the idea

Figure 2.20 International Internet Bandwidth, 2009

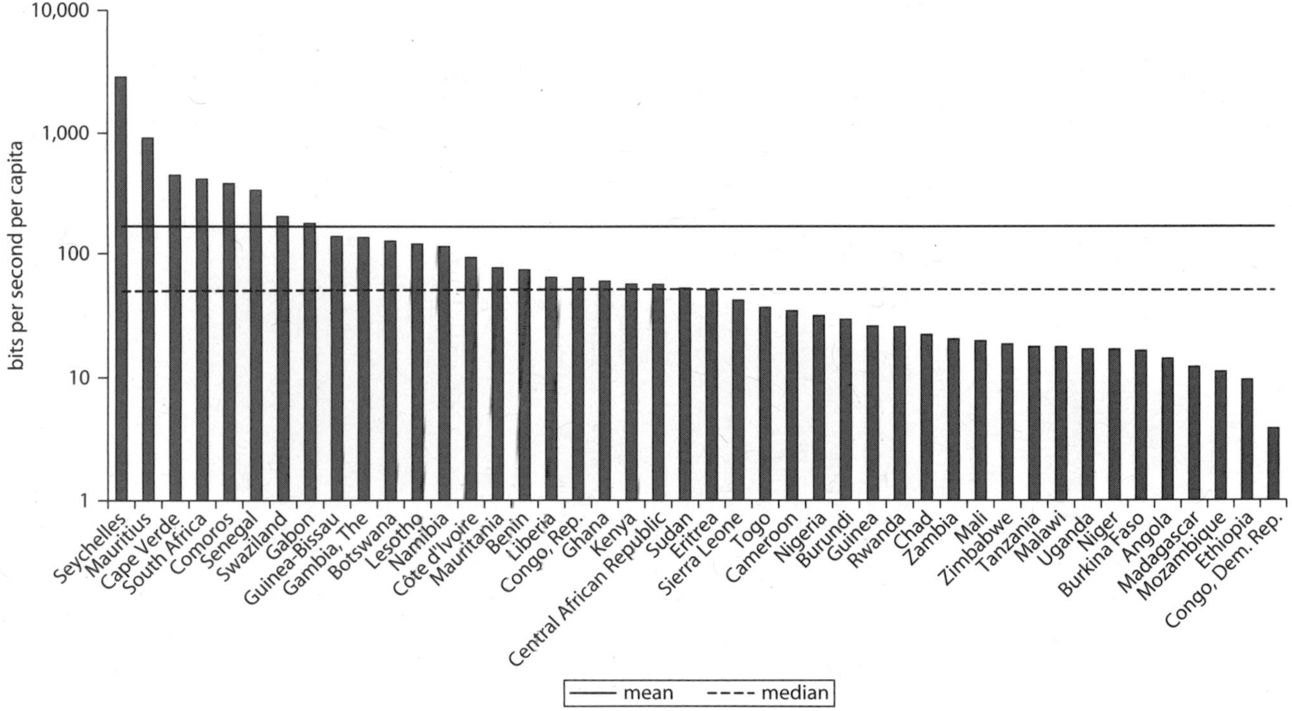

Source: TeleGeography 2010.
Note: Logarithmic scale.

of physically separate networks an accurate representation of the design of modern telecommunications networks. Operators buy and sell network services to one another, in effect using others' networks to fill gaps in their own. In high-income countries, many mobile operators own only the wireless last mile, while piggybacking on other networks for the other segments. In Africa, where networks have traditionally been built as standalone, end-to-end networks, this type of integration is less common.

Other differences are also found between the architecture of networks in Africa and countries with more advanced telecommunications markets. Mobile networks in Africa tend to be wireless throughout, rather than using fiber-optic networks in their core, although this is beginning to change as operators upgrade parts of their networks to fiber-optic cables. Another difference is that some operators in Africa connect to their customers using wireless access networks that do not allow mobility. These "fixed-wireless networks" have typically been used to provide links to businesses and homes in place of copper-based access infrastructure.

These technology-based distinctions are starting to become obsolete as once-distinct technologies converge. Fixed-wireless networks are becoming mobile, wireless networks are being upgraded to fiber, and networks that were once used to provide voice services are increasingly being used to provide a full range of telecommunications services. This process is also being driven by the business side. The purchase of Ghana Telecom by the international mobile operator Vodafone, for example, included fixed and mobile licenses. France Telecom has also bought incumbent fixed-line operators with mobile operations (for example, Côte d'Ivoire Telecom, Kenya Telecom, and Sonatel of Senegal). Finally, licensing regimes across the region are gradually evolving from ones based on the technology to be deployed to more general, non–technology-specific ones. This process is further eroding the traditional distinction between fixed and mobile operators. Despite these changes in the networks and the telecommunications market, however, traditional concepts of fixed and mobile networks remain useful analytically. This section therefore uses the traditional classification of network types in its discussion of the deployment of telecommunications network infrastructure in Africa.

Fixed Networks

The first telephone networks built in Africa were wireline networks, as is the case in other parts of the world. These networks typically connected customers using fixed copper-based lines. Core networks were also traditionally built of copper, but recently, operators have begun replacing these networks with fiber-optic links capable of carrying higher volumes of traffic.

Fixed networks in Africa have not grown significantly in recent years. In large part, this is because of strong competition from mobile networks, but high investment costs, high operating costs, and routine theft of the copper wires are also deterrents. As a result, the fixed-line segment of the market has remained static in most countries in Africa. One notable exception is Nigeria, in which the number of fixed lines has increased—but even this growth has been mainly in fixed wireless links, not traditional copper-wireline–based networks.

For voice calls, mobile networks provide a good technical substitute for traditional fixed networks, so fixed networks' lack of growth need not, in itself, concern policy makers. However, the small size of the fixed-line networks and the poor quality of the infrastructure have implications for the growth of broadband Internet in the region. Broadband in most other parts of the world has been partly driven by the upgrading of copper telephone networks to provide customers with high-speed data services, but because of the shortcomings of these networks in Africa, it was not until wireless broadband technologies became widely available that broadband took off in the region.

Mobile Networks

The rapid growth in access to voice telephony in Africa has been facilitated by investment in mobile networks. The GSM standard has emerged as the dominant one, although there are a few countries in which the other major alternative standard—CDMA—networks is operational.[12] GSM mobile networks have expanded steadily to cover an increasing share of the region's population (figure 2.21)

By 2009, 90 percent of Africa's urban population was living within reach of a mobile network,[13] up from 17 percent a decade earlier. Coverage in rural areas was much less extensive, with about 48 percent of the population living within physical reach of a mobile network, up from just 5 percent a decade ago. In contrast to urban coverage growth rates, which have begun to slow as networks reach full coverage, the rate of growth of rural coverage has remained steady at about 4 percentage points every year (see chapter 5 for a more detailed discussion of mobile coverage).

As with other measures of sector performance, mobile network coverage rates vary widely among African countries. The average population coverage in the top 10 performing countries in Sub-Saharan Africa is nearly five times as great as that of the bottom 10. Rates of coverage in

Figure 2.21 GSM Footprint, 1999 and 2009

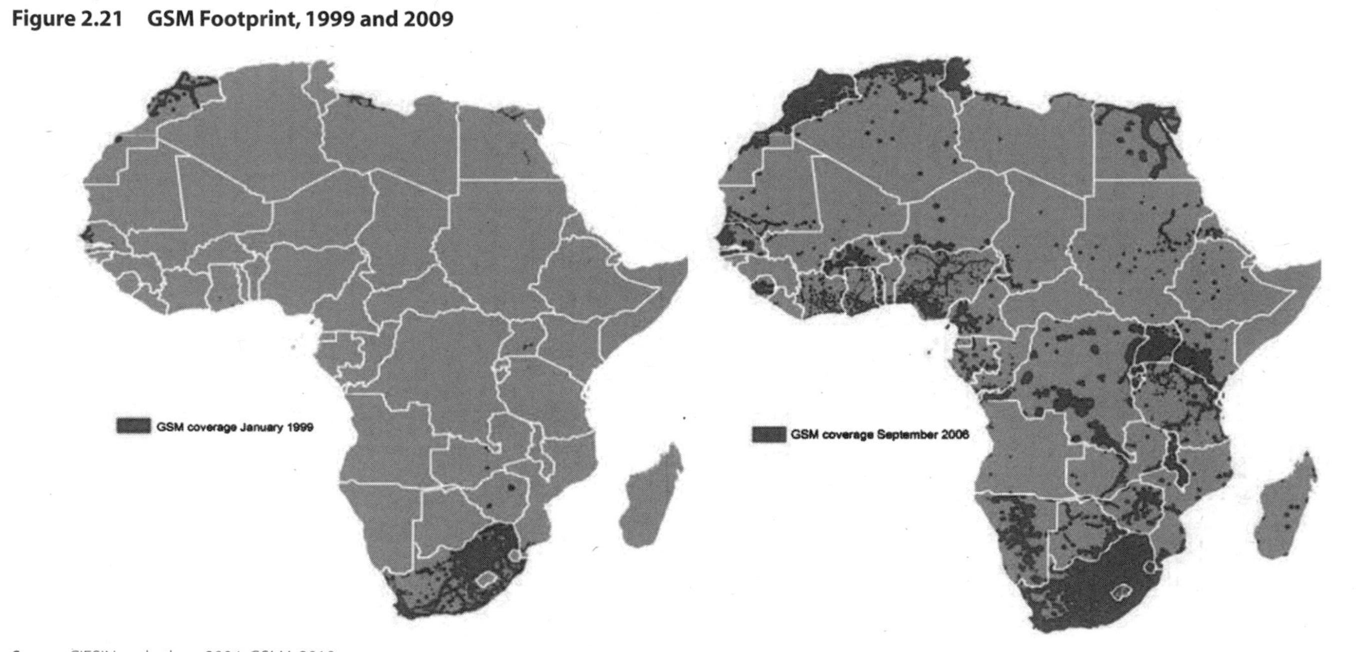

Source: CIESIN and others 2004; GSMA 2010.
Note: Data for some countries are not available.

urban areas also differ. On average, the top 10 countries in Africa had mobile networks covering 95 percent of their rural population, compared with only 5 percent in the bottom 10 countries (figure 2.22).

What explains this variation? Looked at from the regional level, basic patterns of network coverage become clear. Countries with a small land area (such as Cape Verde, the Comoros, Mauritius, the Seychelles, and São Tomé and Príncipe) tend to have higher mobile coverage rates, regardless of the degree of competition, because they are more likely to be richer than average and have higher population densities, requiring fewer base stations to cover the market. Larger African nations, particularly those with low population densities, tend to have lower rates of network coverage. The Democratic Republic of Congo, for example, is a large country with a low population density—28 people per square kilometer (km^2) compared with an average in Sub-Saharan Africa of 35 people per km^2 (World Bank 2010). Accordingly, mobile networks in the country covered only 54 percent of the population in 2009, compared with the average for Sub-Saharan Africa of 61 percent.

Geographic factors alone, however, do not determine the extent of network coverage. Countries with little or no competition among mobile operators have lower rates of population coverage than countries with

Figure 2.22 Mobile Network Population Coverage, 2009

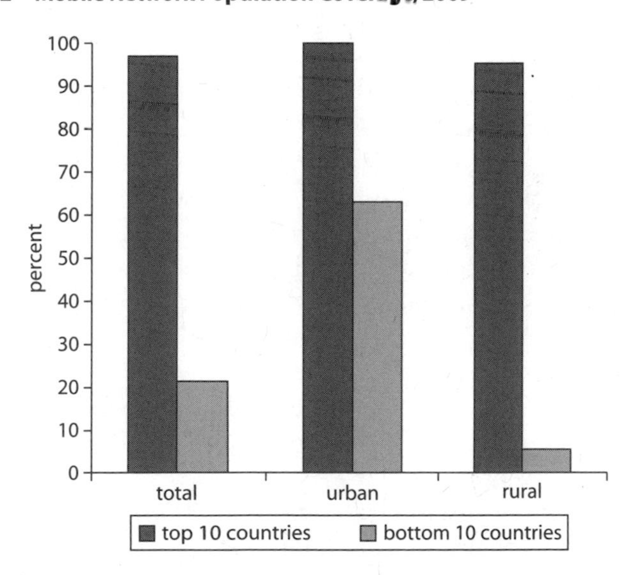

Source: Ampah and others 2009; GSMA 2010.

highly competitive mobile markets, all other things being equal. For example, Ethiopia is a large country with a much higher population density than the Democratic Republic of Congo (81 people per km^2), suggesting that it would have a higher level of coverage, but unlike the Democratic Republic of Congo, Ethiopia maintains a monopoly in the provision of telecommunications services. Largely on account of this, its mobile network covers only 11 percent of the population, less than one-fifth of the regional average.

Broadband Access Network Infrastructure

Many different technologies are used to provide customers with access to broadband Internet, and these technologies have evolved rapidly over time. In 2005, 40 percent of fixed Internet subscriptions in the OECD were dial-up, but by 2008 this figure was down to 10 percent. In some of the more advanced Internet markets, such as Korea, dial-up has practically disappeared (OECD 2009b).

Broadband, which has largely replaced dial-up, was initially provided via copper-based telephone networks that were upgraded to carry broadband data[14] or via upgraded cable television networks. More recently, companies around the world have begun investing in fiber-to-the-home access networks, which are capable of providing fast broadband services.

In parallel with these fixed-line networks, wireless broadband networks have also been evolving and are expanding rapidly. Many different standards for this are seen, including the 3G family of mobile standards,[15] WiMAX, and Long Term Evolution (LTE). No one standard has yet emerged as the global dominant technology. In OECD countries, 18 percent of mobile users have wireless broadband (3G) access (OECD 2009b), but in these countries, the primary means of broadband access remains fixed-line technologies such as DSL (digital subscriber line), cable, or fiber-to-the-home. In high-income countries, wireless broadband is currently seen as a complement to fixed-line access, rather than a substitute.

The path of broadband market growth in Sub-Saharan Africa has been quite different. The lack of high-quality copper telephone lines and the virtual absence of cable television mean that wireline broadband access networks are very limited. Wireless technologies are therefore playing a much more significant role in the provision of broadband in the region than in other parts of the world.

WiMAX was the first widely deployed wireless broadband Internet access network technology in Africa; Internet service providers (ISPs) and fixed-network operators typically adopted it as a fixed-access technology[16] to provide broadband Internet to customers in the absence of a widespread copper access network. By mid-2009, 88 WiMAX networks had been licensed in Africa, of which 35 were commercially operational (TeleGeography 2009).[17] It is estimated that these networks cover approximately 80 million people (WiMAX Forum 2010), but data on the total number of WiMAX subscribers are not currently available.

Mobile broadband access technologies have recently begun growing rapidly as mobile operators have turned their attention to broadband services (*Telecom Finance* 2010) (figure 2.23). Figure 2.23 reveals that at least one of the constraints that has held back broadband in Africa—the lack of access infrastructure capable of delivering broadband Internet—has begun to ease. This trend is likely to persist as the prices of wireless broadband network equipment and computers continue to fall. It will also be further boosted by the commercial launch of new mobile wireless broadband networks using standards such as LTE, which are likely to take place in Africa in the near future (Business Monitor International 2010b).

Figure 2.23 Wireless Broadband Subscribers in Sub-Saharan Africa, 2005–10

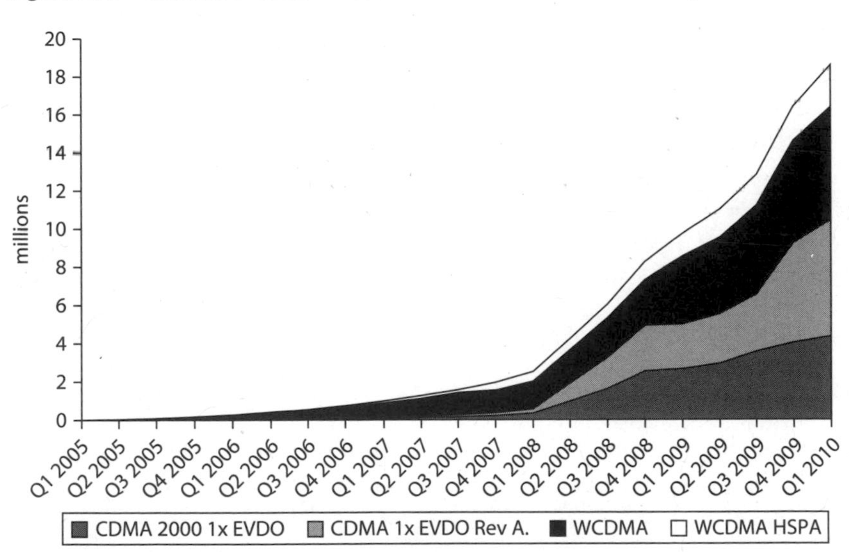

Source: Wireless Intelligence.
Note: Q = quarter. CDMA = Code Division Multiple Access; EVDO = evolution–data optimized; HSPA = high-speed packet access; WCDMA = Wideband Code Division Multiple Access.

Backbone Network Infrastructure

As the number of broadband subscribers increases, so does the amount of traffic being carried on the networks; operators therefore need to invest in the core infrastructure that carries the traffic. Operators have traditionally used fixed networks of copper and, more recently, fiber-optic cable for broadband. Mobile operators in Africa, however, typically use wireless technologies exclusively (Williams 2010). Although wireless backbone networks can be built quickly and relatively inexpensively, they are adequate for only low traffic volumes. Fiber-optic backbones can carry much greater volumes of traffic but are more expensive to build. Still, as traffic increases, the average cost of transporting that traffic falls quickly so that, once a certain volume is reached, fiber-optic networks become more cost efficient than wireless networks.

On any route, the traffic volume and the geographic distance that communications traffic must travel are the most important determinants of which technology will be used. For example, fiber-optic links are installed on high-traffic routes, whereas links carrying lower volumes of traffic often remain wireless. A basic rule for guiding technology selection is given in table 2.1.

Once a certain traffic level is reached, fiber-optic cable becomes the most cost-effective type of communications infrastructure. This is particularly important to keep in mind when considering the ability of Africa's telecommunications infrastructure to provide broadband Internet services. Because these services generate much larger volumes of traffic than voice services (figure 2.24), the extent of a country's fiber-optic infrastructure can become a limiting factor on its ability to deliver high-bandwidth services to customers.

The recent growth of broadband services in Africa has stimulated investment in fiber-optic backbone infrastructure, even by operators that previously had entirely wireless-based networks. Fiber-optic backbone

Table 2.1 Choice of Broadband Backbone Technology Based on Traffic Volume and Distance Traveled

	Capacity		
Distance	*<8 Mbps*	*8–450 Mbps*	*>450 Mbps*
<100 km	Satellite/microwave	Microwave	Fiber optic
>100 km	Satellite	Microwave/fiber optic	Fiber optic

Source: Williams 2010.
Note: Mbps = megabits per second.

Figure 2.24 Backbone Bandwidth Requirements for Various Communication Services

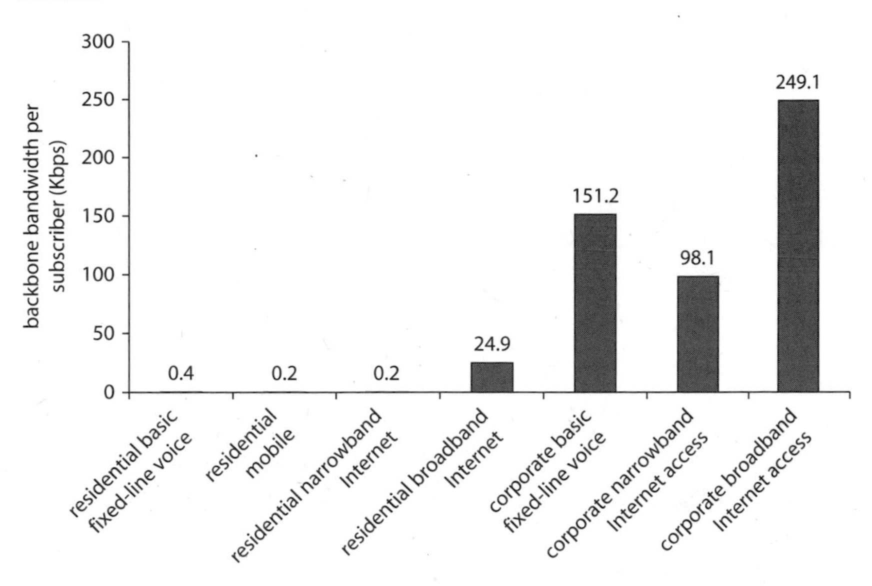

Source: Williams 2010.

networks on the continent are now extensive and expanding rapidly (figure 2.25).

By December 2009, operators in Sub-Saharan Africa had 234,000 km of operational fiber-optic backbone networks. This was not distributed equally throughout the region; a handful of countries account for the majority of the backbone network infrastructure (figure 2.26). South Africa alone accounts for two-thirds of the total, with only 80,000 km in the rest of the region. After South Africa, Nigeria has the second-most-extensive fiber-optic backbone, with about 21,000 km. Despite a low per capita GDP, Nigeria has a large land area and a very large population, approximately half of which live in urban areas (World Bank 2010). As a result of policies that have promoted backbone infrastructure competition, the country also has a very competitive market, which has resulted in multiple fiber-optic networks running along major routes.

Fiber-optic backbone networks across Africa are growing rapidly, with 41,000 km—or 17 percent of the total existing network length—currently under construction. Furthermore, although the rest of the

Figure 2.25 Fiber-Optic Backbone Infrastructure Deployed and Planned, 2009

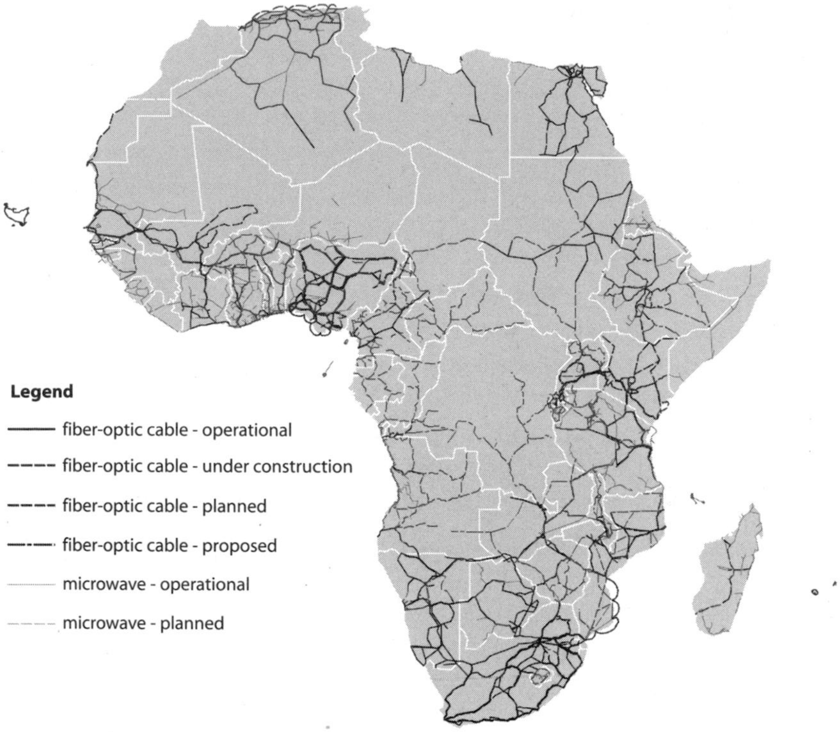

Source: Hamilton 2010. © Hamilton Research Ltd.

region still lags far behind South Africa, 37,000 km of the 41,000 km (90 percent) of the fiber-optic backbone network under construction are in other countries. At this rate of growth, the total stock of fiber-optic backbone network in the region, excluding South Africa, will double in length in less than two years.

This fiber-optic network development also represents a major investment in the ICT sector. It is estimated that the networks that are currently under construction in Sub-Saharan Africa account for about $0.8 billion of investment. To put this into perspective, it is estimated that the total annual investment in the telecommunications sector is about $5 billion per year (see chapter 4). Current investment in fiber-optic backbone infrastructure in Sub-Saharan Africa is therefore equivalent to over 15 percent of the total annual investment in the sector. This is an indication of the higher levels of traffic arising from rapidly growing

Figure 2.26 Length of Operational Fiber-Optic Backbone Networks in Top 10 Countries in Sub-Saharan Africa (excluding South Africa), End 2009

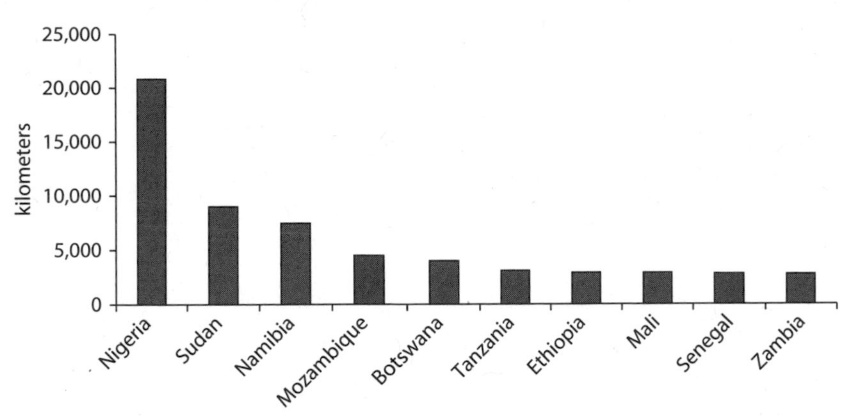

Source: Hamilton 2010.

voice customers, as well as the increasing importance of broadband in the sector.

The rate of network expansion varies widely among countries in the region. The fiber-optic backbone is growing most quickly in Kenya, followed closely by Nigeria (figure 2.27). Kenya's rate of expansion is especially remarkable given its starting point. Nearly 7,000 km of fiber-optic backbone networks are under construction in Kenya—approximately four times the current length of operational backbone networks. Similar lengths are under construction in Nigeria, although its network is expanding at a proportionately slower rate, given that the existing network is much larger. Another interesting point to note is the variation in countries' network growth rates, in both absolute and relative terms. Kenya and Ghana, for example, are of similar size, but Kenya's networks are growing much more quickly—with 6,445 km versus Ghana's 919 km of backbone network currently under construction. Figure 2.27 also shows that small countries, such as Rwanda, are equally able to experience high rates of growth in their fiber-optic backbone networks.

Are such high growth rates sustainable? At this point it is too early to predict how the market for fiber-optic networks will develop. Nevertheless, a further 86,000 km of network infrastructure in Sub-Saharan Africa have either been proposed or are in the planning stage, which indicates that these networks are going to continue growing, at least in the short term.

Figure 2.27 Length of Fiber-Optic Backbone Network under Construction in Top 10 Countries in Sub-Saharan Africa (excluding South Africa), End 2009

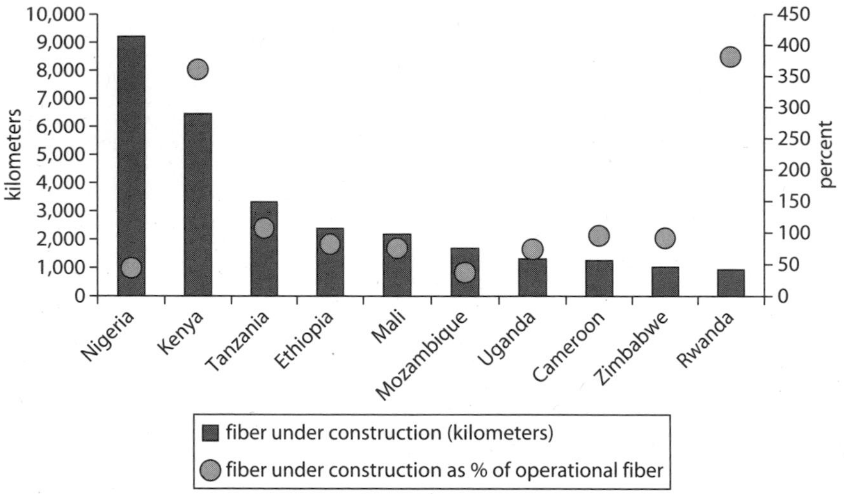

Source: Hamilton 2010; World Bank staff analysis.

Submarine Fiber-Optic Infrastructure

The submarine fiber-optic network infrastructure is an essential complement to terrestrial backbone networks. It provides a high-bandwidth, low-cost alternative to satellites for carrying traffic to and from the Africa region. The development of this infrastructure is therefore crucial for the provision of affordable broadband services. Until recently, however, the region has remained largely unconnected to the global submarine cable networks. South Africa has been the exception to this since 1967, when the South Atlantic 1 (SAT-1) cable (now decommissioned) was launched connecting the country to Ascension Island, followed by SAT-2 in 1993 connecting South Africa to Europe. In 2002, a new submarine cable system (the South Atlantic 3/West Africa Submarine Cable, more commonly known by its acronym SAT-3/WASC) entered service, connecting South Africa and countries along the west coast of Africa to Europe. Until 2009, the only other countries with access to submarine cables were Cape Verde, Djibouti, and Mauritius. Since then, the situation has changed dramatically. By 2010, Sub-Saharan Africa had 12 operational cables, and another 5 were under construction. The operational cables have a combined capacity of over 12 terabits per second (Tbps). A total of $1.7 billion is being invested in the five submarine cables currently under construction, which will bring an additional 9 Tbps of capacity to the region. Africa's submarine fiber-optic cable infrastructure is illustrated in figure 2.28 (see appendix table A1.1 for details).

Investing in submarine cables has significant implications for the region. Satellites require an investment of between approximately $250 million and $650 million, depending on their size and payload. The next generation of satellites under construction using Ka-band frequencies has much greater data transmission capacities than the older generation. Eutelsat's KA-SAT, launched over Europe in 2010, has a capacity of 70 Gbps; Viasat-1, launched over North America, has over 100 Gbps.

Figure 2.28 Submarine Fiber-Optic Cables in Africa, 2011

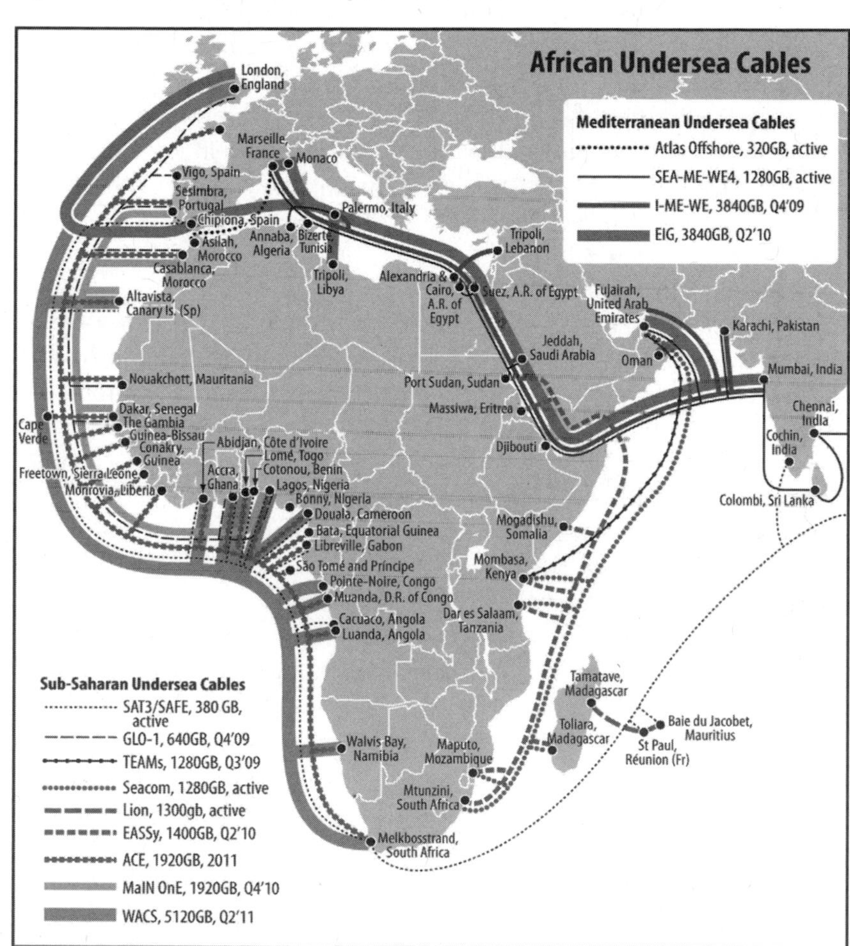

Source: Hamilton 2010. © Hamilton Research Ltd.

Meanwhile, O3b Networks has announced the launch of two constellations of eight satellites, each with a global capacity of 10 Gbps, providing coverage all over Africa. Yet the capacity of satellites remains small in comparison to that of submarine fiber-optic cables. The SEACOM submarine cable, for example, which entered service in July 2009, cost approximately $650 million and has a design capacity of 1.28 Tbps. The Eastern Africa Submarine Cable System (EASSy), which entered into service in 2010, cost $235 million and has a maximum design capacity of 1.4 Tbps.

The total capacity available is not the only factor determining the impact of submarine cable infrastructure on the broadband market. The level of competition among cables is also a key factor (see chapter 3 for a more detailed discussion of this). The other factor is the extent of the terrestrial backbone networks that are used to carry traffic to and from the submarine cable landing stations. Without extensive terrestrial backbone networks, the impact of the submarine cables on the broadband market might be limited. Satellites, however, can deliver their lower capacity directly to the customer. Satellites are therefore likely to play a significant role in Africa's broadband landscape for the foreseeable future.

Internet Exchange Points

Most data traffic generated by users in Africa passes in and out of the region,[18] in part because much of the Internet content is stored outside of the region. However, the absence of infrastructure to facilitate the exchange of traffic among local ISPs (that is, within Africa) has meant that even traffic that is passing from one African Internet user to another often passes outside of the region and is exchanged by ISPs in Europe or North America. The historically high cost of international bandwidth in Africa means that this international routing of intra-African traffic has imposed a cost on Internet users in the region.

One solution to this is to establish Internet exchange points (IXPs) that route intra-African traffic, thereby avoiding unnecessary international transit. Several different models are found for establishing and governing IXPs. For example, some are private, for-profit entities that charge ISPs to interconnect and exchange their traffic; others are nonprofit, cooperatively managed institutions under which ISPs exchange traffic among themselves, usually at low or no cost.

The experience of IXPs in Africa has been mixed. Many attempts have been made to establish them, but only a few have been successful, as measured by the number of ISPs connected and the volumes of traffic that they carry. In Ghana, for example, there are two IXPs: the Ghana Internet Exchange (GIX), established in 2005, and the Accra Internet Exchange (AIX), also established in 2005. Kenya, however, has been more successful in this area. The Kenya Internet Exchange Point (KIXP) was established in 2002,[19] and by 2008 had 26 members and was routing 36 Mbps of traffic.

Two key issues appear to be driving the success—or failure—of IXPs in Africa. The first is the basic economics of such facilities. Many different factors affect the commercial viability of an IXP. One is the cost associated with designing, building, and operating the facility. If the money saved by exchanging traffic locally is insufficient to cover this cost, the facility will likely fail. In countries with very low Internet subscriber bases or where most Internet subscribers use narrowband connections, it is unlikely that IXPs will be economically feasible. As traffic volumes increase, however, the economics of IXPs change. On the one hand, higher volumes mean greater potential cost savings from switching traffic in the IXP rather than routing it outside of the region, but on the other hand, at the same time that the number of broadband subscribers has increased, traffic patterns have also changed. Modern Internet usage creates data traffic patterns that consist less of e-mail exchange, which is mainly intracountry, and more of the downloading of web pages and media content, most of which is stored outside the region. This effect reduces the need for a locally based IXP. The other force affecting the economics of IXPs is the cost of local and international bandwidth. As international bandwidth prices fall, the economic rationale for establishing an IXP is reduced. This is particularly true in countries where domestic bandwidth (that is, backbone services or backhaul) is expensive. In these countries, it may be cheaper to route traffic through an exchange point in Europe than it is to route to an IXP over terrestrial networks.

The second major factor driving the success of IXPs in Africa has been their organization. As ISPs compete with one another, they can find it difficult to set up the cooperative management arrangements that are required for a well-functioning IXP. Large ISPs (known as Tier 1 ISPs) would sometimes rather connect directly to one another and exchange traffic by way of "peering" relationships rather than through a common IXP serving the entire industry. In South Africa, for example, the major ISPs connect to one another through peering relationships, whereas the small, Tier 2 ISPs connect at the Johannesburg Internet Exchange point, which then is connected to the Tier 1 players.

15. These include CDMA2000 1xEVDO, CDMA2000 1xEVDO Rev.A, WCDMA, and WCDMA HSPA.
16. The first generation of WiMAX broadband access technologies that was widely deployed in Africa (IEEE 802.16d) was a fixed connection. Subsequent WiMAX standards (IEEE 802.16e and later) are designed to allow mobility. These later generations of WiMAX networks are not yet common in Africa.
17. WiMAX figures include countries in North Africa.
18. Much of the information in this section comes from Amega-Selorm and others (2009).
19. It was initially launched in 2000 but experienced difficulties getting regulatory approvals. It was relaunched in 2002.

References

Adegoke, A. S., I. T. Babalola, and W. A. Balogun. 2008. "Performance Evaluation of GSM Mobile System in Nigeria." *Pacific Journal of Science and Technology* 9(2): 436–41.

Amega-Selorm, Charles, Muriuki Mureithi, Dobek Pater, and Russell Southwood. 2009. "Impact of IXPs—A Review of the Experiences of Ghana, Kenya and South Africa." Open Society Institute, London.

Ampah, Mavis, Daniel Camos, Cecilia Briceño-Garmendia, Michael Minges, Maria Shkaratan, and Mark Williams. 2009. "Information and Communications Technology in Sub-Saharan Africa: A Sector Review." AICD Background Paper 10, Africa Infrastructure Country Diagnostic, World Bank, Washington, DC.

ARTP (Agence de Régulation des Télécommunications et des Postes). 2007. "Enquête Qualité de Service des Réseaux GSM au Sénégal–Octobre-Novembre 2006." ARTP, Dakar.

Business Monitor International. 2010a. "Nigeria Telecommunications Report Q42010." Business Monitor International, London.

———. 2010b. "Kenya: Safaricom to Begin LTE Technical Trials." Business Monitor International, London.

CIESIN (Center for International Earth Science Information Network, Columbia University), IFPRI (International Food Policy Research Institute), World Bank, and CIAT (Centro Internacional de Agricultura Tropical). 2004. *Global Rural-Urban Mapping Project (GRUMP): Urban/Rural Population Grids.* Palisades, NY: CIESIN, Columbia University. http://sedac.ciesin.columbia.edu/gpw.

Communications Commission of Kenya. 2009. "Annual Report 2007–2008." Communications Commission of Kenya, Nairobi. http://www.cck.go.ke/resc/publications.html.

FCC (Federal Communications Commission). 2010. *Sixth Broadband Deployment Report*. Washington, DC: FCC.

Foster, Vivien, and Jevgenijs Steinbuks. 2009. "Paying the Price for Unreliable Power Supplies: In-House Generation of Electricity by Firms in Africa." Policy Research Working Paper 4913, World Bank, Washington, DC.

GSMA (GSM Association). 2007. *Global Mobile Tax Review 2006–2007*. London: GSMA.

———. 2010. Untitled data file on GSM mobile network location provided by GSMA. GSMA, London.

Hamilton, Paul. 2010. "Broadband Network Development in Sub-Saharan Africa." Unpublished paper, Hamilton Research, Bath, England.

ITU (International Telecommunications Union). 2009. *Measuring the Information Society—The ICT Development Index*. Geneva: ITU.

———. 2010. *Measuring the Information Society 2010*. Geneva: ITU.

Lange, Peter. 2010. *African Mobile Broadband, Data and Mobile Media Market*. Australia: Paul Budde Communications.

Lewis, Charley. 2010. "Achieving Universal Service in South Africa: What Next for Regulation?" Conference paper, International Telecommunications Society Asia-Pacific Regional Conference on Telecommunications Ubiquity and Equity in a Broadband Environment." August 26–28, 2010. Wellington, New Zealand.

Millicom International Cellular SA. 2007. "Form 20-F." Millicom International Cellular SA, Luxembourg.

OECD (Organisation for Economic Co-operation and Development). 2002. "OECD Mobile Basket Revision, Working Party on Telecommunication and Information Services Policies." OECD, Paris.

———. 2005. *OECD Communications Outlook 2005*. Paris: OECD.

———. 2008. *Broadband Growth and Policies in OECD Countries*. Paris: OECD.

———. 2009a. "OECD Broadband Statistics 2009." OECD, Paris. http://www.oecd.org/document/54/0,3746,en_2649_33703_38690102_1_1_1_1,00.html.

———. 2009b. *OECD Communications Outlook 2009*. Paris: OECD.

Qiang, Christine Zhen-Wei, and C. Rossotto. 2009. *Economic Impacts of Broadband in 2009 Information and Communications for Development*. Washington, DC: World Bank.

Telecom Finance. 2010. "Africa: A Telecom Myth or a Continent of Genuine Opportunity?" *Telecom Finance* (London) no. 177. March.

TeleGeography. 2009. "WiMAX Market Review Q2 2009." PriMetrica, Washington, DC.

———. 2010. "TeleGeography's Bandwidth Pricing Report." PriMetrica, Washington, DC.

UCC (Uganda Communications Commission). 2008. "Status of the Communications Market—September 2008." UCC, Kampala.

Williams, Mark D. J. 2010. *Broadband for Africa: Developing Backbone Communications Networks*. Washington, DC: World Bank.

WiMAX Forum. 2010. "Industry Research Report." WiMAX Forum, Portland, OR, May.

Wireless Intelligence (database). http://www.wirelessintelligence.com.

World Bank. 2010. *World Development Indicators.* Washington, DC: World Bank.

World Bank Development Data Platform. http://databank.worldbank.org/ddp/home.do.

CHAPTER 3

Market Reform and Regulation

The rapid growth in Africa's telecommunications infrastructure and access to telecommunications services, described in chapter 2, has been associated with the reform of telecommunications markets throughout the region. Since the late 1990s, almost all countries have liberalized their telecommunications sector, and competition has developed rapidly as a result. But liberalization has not proceeded at the same pace across all market segments. By 2009, competition among mobile operators had been introduced in 42 countries in Sub-Saharan Africa. As many as five operators are now found in some countries, indicating that highly competitive market structures are possible even in low-income countries. However, fixed-line telecommunications remain uncompetitive in many countries, and in some, a state-owned operator retains the monopoly. Yet the few countries that have fully liberalized the fixed-line segment of their market, such as Nigeria, have experienced rapid growth, especially where wireless operators offer both voice and data services in competition with the incumbent fixed-line operator.

Another area that has not undergone as much change as the mobile segment of the market is backbone infrastructure. Governments continue to constrain investment in this area either through the outright support or control of a monopolistic operator or through regulatory and licensing restrictions that make it unprofitable for companies to develop backbone networks. Yet competition is feasible and profitable here, too, as evident in countries—such as Kenya, Nigeria, and Sudan—where backbone operators have been allowed entry and have established networks.

The structure of ownership of telecommunications operators is evolving as the liberalization process continues. One important trend is the emergence of regional operators with operations in multiple countries across Africa. This allows operators to benefit from economies of scale when purchasing network equipment and is driving regional network integration. Meanwhile, incumbent operators are being privatized across the region—though this process is far from complete.

Liberalization and privatization, on their own, are not sufficient to develop a competitive telecommunications sector. Institutional reforms are also needed to ensure effective competition. The most important of these reforms are the establishment of regulatory authorities and the development of regulatory frameworks that support competition. Governments throughout Africa have established authorities to regulate their telecommunications markets. Most of these have some degree of autonomy, which is necessary for them to make balanced and technical regulatory decisions without undue political influence or government pressure. Yet the independence of regulators and the subsequent quality of their decisions are far from uniform—variations that are in part reflected in the extent of competition in markets across the region.

Major regulatory measures that support competition in the telecommunications market include control over interconnection charges, the establishment of mobile virtual network operators, and the introduction of portable mobile numbers. Only a few regulators have implemented all of these measures, although momentum to do so seems to be building across the region. Another important responsibility that often falls to regulators is the establishment and operation of universal service funds, which center on contributions from operators and are supposed to be used to increase access to information and communication technology (ICT) in under- and unserved areas. These funds have a mixed track record, however, and many remain undisbursed. In practice, the most effective promoter of universal services has been market liberalization, which has dramatically widened network coverage and reduced prices.

The profound reforms in sector structure and regulation in Africa's telecommunications market have coincided with major improvements in sector performance. Privatization and increased competition are associated with greater revenues and higher subscriber penetration rates. The impact on prices, however, is less clear. Prices can rise immediately following liberalization as new operators supply previously unmet demand. As competition develops, however, prices begin to fall.

Market Liberalization and the Development of Competition

Most countries in Sub-Saharan Africa have introduced some degree of competition into their telecommunications markets. By 2009, all but four countries in the region (Comoros,[1] Ethiopia, São Tomé and Príncipe, and Swaziland) had opened their mobile markets to competition. One other country, Eritrea, had passed legislation to introduce competition but had not implemented the policy by issuing additional licenses. About half of the countries in the region had allowed competition in the market's fixed-line and international segments (table 3.1). The provision of Internet access had also been liberalized in most countries. Basic information on the national telecommunications legislation in force in the countries of Sub-Saharan Africa is reported in appendix table A2.1.

Only about one-third of the countries in the region have more than one local fixed-line operator, and barely half have more than three mobile operators. The downstream (retail) market for Internet services is usually competitive, but governments often limit competition in upstream segments such as domestic backbone networks or access to international submarine cables. Market liberalization is therefore often more advanced on paper than in practice.

This gap between legislation and practice is particularly evident in the licensing process. For example, in Namibia legislation provides for competition in fixed-line and international gateway facilities, but delays in issuing licenses to new entrants have hindered the development of competition. The cost of licenses can also present a significant barrier to potential market entrants. The fees involved, when combined with other regulatory charges—such as universal service contributions and spectrum usage fees—can make market entry prohibitively expensive. In Zambia, for example, at the end of 2009 the international facilities segment of the market was, in theory, open to competition, but the price of an international voice gateway license was as high as $12 million. In contrast, a public infrastructure provider license has a one-time entry fee of just $100,000 in Uganda.[2] This explains why, by 2009, a competitive international services market had developed in Uganda whereas in Zambia, there was still only one international gateway operator—the state-owned Zambia Telecommunications Company Ltd. (Zamtel).[3] According to the

Table 3.1 Status of Telecommunications Market Competition, Sub-Saharan Africa, 2009

	Legal status of competition				*Number of operators*			
	Monopoly	*Partial competition*	*Competition*	*Data not available*	*1*	*2*	*>2*	*n.a.*
Mobile	4	14	26	3	6	15	26	
International	18	7	17	5	*	*	*	*
Internet	4	4	34	5	3	4	35	5
Local fixed	17	7	16	7	32	9	6	1
International gateway	10	11	12	14	a	a	a	a

Sources: Adapted from the International Telecommunications Union (ITU) and regulator websites.
Note: The table shows the number of countries in each category (that is, in four countries, the legal status of the mobile operator is that of a monopoly).
a. Because of problems with definitions regarding the number of mobile operators with direct international voice connectivity and Internet service providers (ISPs) with direct international gateway access, it is not possible to compile the number of operators for these categories.

United Nations Conference on Trade and Development (UNCTAD 2007, p. 15):

> International call costs in Zambia are among the highest in the region, not all connections (incoming and outgoing) are successful and calls are often of poor quality. This has been frequently cited by investors as contributing to the high cost of doing business in the country. All international calls are currently routed through an international gateway operated by ZAMTEL. However, this gateway is unable to provide for the required traffic because of a lack of investment in equipment and the fact that ZAMTEL has no competition which could provide the incentive to do so.

Meanwhile, sudden changes to license fees increased market risk and uncertainty, creating additional barriers to entry. Many cases are seen in which governments or regulators have introduced significant changes to the charges imposed on operators for their licenses. In 2007, for example, the government of Benin increased license fees and imposed the new rates retroactively, ordering operators to pay fees of about $50 million in addition to the fees originally agreed upon at the time of license award (Global Insight 2007).

As table 3.1 shows, there has been an overall move toward market liberalization throughout the telecommunications market in Africa, but countries have moved at different speeds, and competition has been introduced into some market segments faster than in others. Mobile markets were liberalized early on, but fixed-line and international gateways opened more gradually. Internet services are a mixed bag. Most countries introduced competition in the downstream parts of this market, allowing Internet service providers (ISPs) to build some of their own wireless infrastructure but also requiring them to use the incumbent operator's fixed-line network. In countries where ISPs have always been allowed to provide data services such as Internet, they have often been restricted from providing voice services via the Voice over Internet Protocol (VoIP) (appendix table A2.2). For a full picture of the liberalization process in Africa, it is necessary to look at each of the major market segments separately.

Competition in Fixed-Line Markets

The fixed-line market is one of the least competitive in the region's ICT sector.[4] This is partly because of the exclusivity periods granted to incumbent operators, but even where such periods have expired or never existed, effective competition has not always developed. Among those countries in which exclusivity arrangements have ceased for at least two years, only half have seen new operators.[5] For example, Telkom South Africa's exclusivity ended in May 2002, but a second national operator was not issued a license until December 2005 and did not commercially launch until August 2006. Despite the lack of formal exclusivity arrangements in Namibia and Zambia, the incumbents there remained the only fixed-line operators long after the official liberalization of their markets.

The growth of mobile operators has played an important role in the stagnation of the fixed-line segment of the market. Mobile phone services in Africa are typically no more expensive than fixed-line services and are, in many cases, much cheaper. Mobile phones are obviously more flexible and, under a prepayment plan, also allow easier control over expenditures, making mobile operators strong competitors to fixed-line operators.

Box 3.1

Fixed-Line Services in Nigeria

Nigeria has been a leader in the process of fixed-line market liberalization and has seen the market segment grow and competition develop—by 2010 it had 30 fixed wireless and independent regional operators. The new system of unified licenses also allowed these operators to provide full-mobility services. But, as this market segment has grown, the market share of Nigerian Telecommunications Limited (NITEL), the state-owned operator, has fallen. By the end of 2009, it was estimated that NITEL had about 50,000 operational lines, approximately 4 percent of the total number of fixed lines in the country.

Regulatory decisions have played an important part in the growth of these fixed-line operators. The government liberalized fixed-wireless market entry with licenses widely available at a reasonable cost. At the same time, growth in the global fixed-wireless market reduced the cost of network equipment and handsets. Restrictions on mobility were lifted in 2003, making the licenses more attractive to investors (*Mobile Leader* 2008). A key driver of investment in the fixed-line segment of the market has been the demand for data services. Before 2006, mobile operators were not providing broadband data services (AfricaNext 2008), leaving a gap in the market that fixed-line operators filled. Finally, delays in privatizing the incumbent operator weakened its ability to compete with the new market entrants.

Source: Authors.

The stagnation of fixed-line networks has also been due to regulatory and policy decisions. Exclusivity agreements have been written into some fixed-line concessions and licenses. Although these can be modified, such a move can often result in legal action and, therefore, is rarely made.[6] Some countries have licensing and regulatory regimes that discourage new market entrants with high and unpredictable license fees, even where market entry is theoretically possible.

One exception to the dominance of the mobile over the fixed-line segment of the market has been in the area of data services. New fixed-line operators have often entered the market with the primary objective of providing data services to customers; meanwhile, mobile networks have concentrated on voice services.

Elsewhere in the region, fixed-line networks have seen steady, if unimpressive, growth over the decade that ended in 2008 (figure 3.1). In some countries, the number of fixed lines has actually gone down. In Sudan, for example, the number of fixed-line subscribers peaked in 2004 at just over 1 million, but this fell by two-thirds over the following four years as customers rapidly shifted away from fixed-line services after the introduction of mobile competition. Similarly, in South Africa, the number of fixed lines fell steadily from a peak of 5.5 million in 1999 to 4.4 million in 2008, despite the introduction of a prepaid pricing platform that made it easier for low-income users to qualify for telephone services.

The use of fixed-wireless technologies to provide broadband data services and lower average network costs, particularly in rural areas, has improved the prospects for the fixed-line market. When used at lower frequencies (for example, 450 megahertz), wireless local loop (WLL) systems have wide transmission abilities suitable for rural and remote areas. According to the CDMA Development Group, more than 30 African countries had commercially deployed a CDMA2000 1× wireless network as of mid-2007. To attract customers, some operators have included additional features, such as limited mobility and free on-net calls. The WLL networks often come with billing platforms that support both contract and prepaid schemes, thereby gaining some of the commercial advantages traditionally enjoyed by mobile networks, but the use of fixed-wireless networks can create challenges for the regulatory authority. In Namibia, for example, the two GSM operators complained to the regulatory authority about the mobility features of the incumbent's fixed-wireless service. Restrictions on this mobility service were lifted as a result of a cabinet decision on May 12, 2009.[7] In Sudan,

Figure 3.1 Net Change in Fixed Lines, Sub-Saharan Africa, 1999–2008

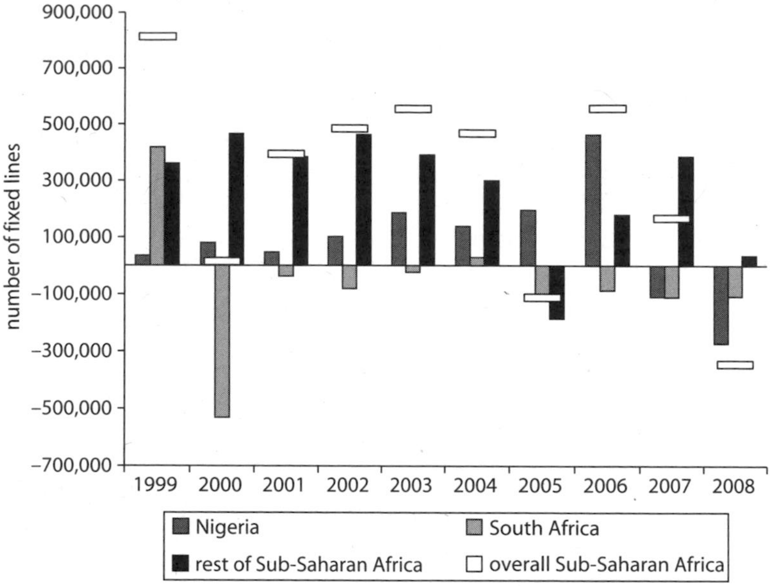

Source: Ampah and others 2009.

Canar, the second fixed-line operator, has lobbied the government since its launch to have mobility restrictions on its fixed-wireless network lifted.

Recently, new communications technologies have emerged to compete with fixed-line operators. Voice services provided over broadband Internet connections are the most important example. In countries where VoIP services are allowed, operators with Internet licenses can provide voice services that compete directly with traditional fixed-line operators. In South Africa, for example, operators with Internet licenses have set up Wi-Fi networks and offer VoIP-based telephone services to the public. In other countries, WiMAX[8] operators now offer voice telephony services to compete with traditional fixed-line operators. Some governments have attempted to prevent VoIP because it impacts the revenue of incumbent operators.[9] Enforcing such restrictions can be difficult, particularly if there is international data gateway competition, because it requires ongoing monitoring of Internet traffic with special equipment that most Sub-Saharan African countries do not have. (More on this on page 83.)

Competition in Mobile Markets

Competition in the mobile segment of the market has emerged steadily as governments have issued more licenses. By 2009, all countries in Sub-Saharan Africa had at least one mobile network, 42 countries had more than one active operator, and more than half had three or more active operators. By comparison, in 1993 three-quarters of countries in the region had no mobile network, and those with mobile networks functioned as monopolies (figure 3.2).

The structure of the mobile market also affects the effectiveness of competition. Some countries have issued multiple mobile licenses, yet their mobile markets remain dominated by one or two major operators; others have a more equal distribution of market shares. One measure of market concentration is the Herfindahl-Hirschman Index (HHI).[10] With five GSM mobile operators and an HHI of 3,414, Nigeria had one of the least concentrated markets in Sub-Saharan Africa in 2009,[11] in part because the government awarded GSM licenses to three operators simultaneously (a fourth license was issued

Figure 3.2 Mobile Voice Market Liberalization in Sub-Saharan Africa, 1993–2009

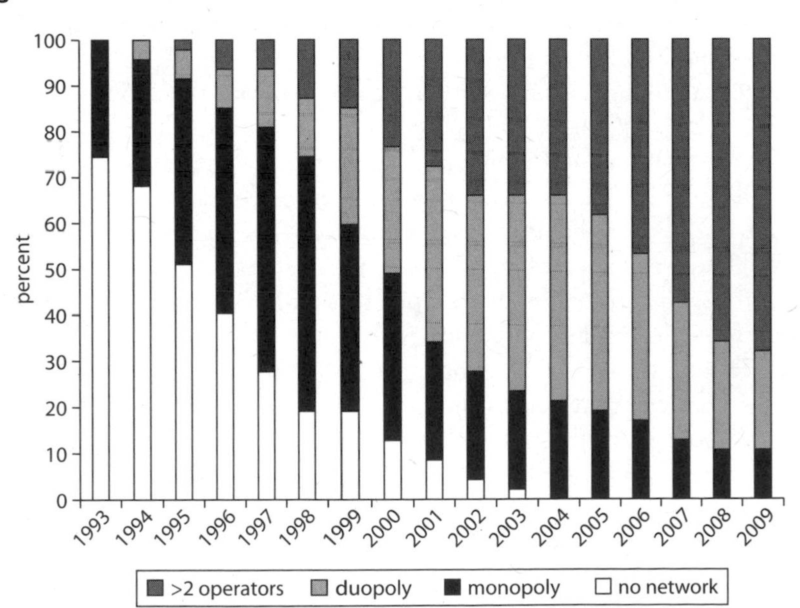

Sources: ITU (2010), regulators, operators.

several years later), thereby contributing to a level playing field. Although the incumbent was awarded one of the licenses, its market position was hampered by an antiquated analog mobile system that had few subscribers and insufficient capacity. Other countries have issued multiple licenses, but competition has developed more slowly, resulting in a more concentrated market. In 2009, for example, Zambia had the same number of operators as Burkina Faso and South Africa, but its market was more concentrated, and it therefore had a higher HHI score (appendix table A2.4).

The evolution of competition in the mobile market has been influenced by the way licenses have been awarded. In Kenya, Lesotho, Mozambique, Senegal, and Sudan, incumbent operators were given several years' lead time through exclusivity agreements before competition was introduced. As a result, later entrants in their markets sometimes struggled to gain market share as rapidly as they might have done otherwise because the incumbent had time to prepare for competition, but this is not always the case. Late entrants have been able to successfully gain market share in some countries, often where they are part of a multinational group or where the incumbent has been slow to develop its network. Examples include the former Areeba (now MTN) in Benin and Ghana; Celtel (sold to Zain and then to Bharti Airtel) in Burkina Faso, Ghana, and Zambia; and Vodacom in Tanzania. Overall, data show that the increase in mobile subscriptions accelerates with the addition of each new operator (appendix table A2.5).

The establishment of effective competition in the mobile market requires more than just issuing licenses. The experience of developed countries indicates that mobile competition can be significantly enhanced through the regulation of mobile termination rates (MTRs), the establishment of mobile virtual network operators (MVNOs), and the introduction of mobile number portability (MNP).

MTRs are the charges that operators levy on one another when a call passes from one network to another. These charges are typically passed on to customers in the prices that they pay for an "off-net" call (that is, when they call a subscriber on a different network). The level of MTRs therefore has an impact on the overall level of retail prices in a market. This may also feed through into the structure of retail prices by pushing up the price of off-net calls relative to on-net calls (see chapter 2). The impact of MTRs on competition is more difficult to determine. It is argued by some that large operators use high MTRs to drive up on-net/off-net price differentials with the aim of giving themselves a competitive advantage over smaller operators trying to enter the market (because their customers would be likely to make more off-net calls than those on large networks). But it is worth noting that the real impact of MTRs on competition is a subject of debate among policy makers, operators, and academics. Some parties have argued that high on-net/off-net price differentials have no impact on competition (Hoernig 2006).

It is possible that competitive strategy does affect operators' decisions on interconnection rates. In South Africa, for example, mobile interconnection rates increased by 500 percent over the three years before the launch of the third mobile operator, Cell C (Esselaar and others 2010). In Kenya, one of the smaller mobile operators complained to the national regulatory authority (NRA) in 2007 about the practice of the dominant mobile operator of charging much lower prices for calls made within its network than for calls made to other networks:

> Early in the year, Celtel wrote to the CCK complaining of alleged monopolistic practices by Safaricom, including the locking in of subscribers through high charges to other networks. Safaricom currently charges its subscribers up to Ksh50 ($0.71) per minute to access the Celtel network, and Ksh45 ($0.64) a minute for calls to Telkom. In contrast, calls terminating within the network are charged as little as Ksh8 ($0.11) per minute. On its part, Celtel charges as low as Ksh16 ($0.22) per minute to call other networks. (*The East African* 2007)

The regulation of MTRs in Africa is discussed in more detail in the section on regulatory issues.

MVNOs are "virtual" operators that establish a brand and a retail business but use the network of another mobile operator. This enhances competition at the retail level but does not significantly affect competition at the infrastructure level. MNP allows customers to transfer their phone number to a new network if they decide to leave their old provider. This eliminates an important barrier to customer switching, which enhances competition among operators.

African regulators are gradually introducing more regulatory controls on MTRs, but most African countries have not yet adopted the other regulatory measures aimed at promoting competition. One major exception is South Africa: Although it has not yet introduced the formal regulatory control of MTRs, it has introduced both MVNOs and MNP. Virgin Mobile was launched as an MVNO in 2005 using the infrastructure of Cell C, one of the three licensed network operators, and MNP was launched in November 2006.

Competition in International Services

Providing international calls has traditionally been a very profitable service for incumbent fixed-line operators, with prices typically well above costs. Where full competition has been introduced, competing operators usually opt to build international network connections to carry their own international calls. This is referred to as "facilities-based" competition, and, for mobile operators, the international facilities concerned have been typically satellite-based links to traffic hubs usually in Europe. Recent investment in fiber-optic submarine cables, described in chapter 2, has the potential to further accelerate the development of international-facilities–based competition.

Full liberalization of international gateways has resulted in intense competition among mobile operators in the provision of international voice calls, which has brought down international call prices significantly (chapter 2). Yet, despite these apparent benefits, some countries in Africa have not fully liberalized this market segment. In these countries, operators are allowed to offer international call services (incoming and outgoing) to their customers, but they are required to use a single network (usually the state-owned incumbent's) to carry the traffic in and out of the country. By 2009, 24 countries in Sub-Saharan Africa reported having established the legal conditions for competition, but, in practice, competition has not been fully established in some of them, and 18 countries retain a monopoly in international communications services.

Regulatory control over the market for international voice services has been complicated by the introduction of competition in the provision of Internet services. ISPs are usually given licenses to carry data traffic in and out of the country to enable them to provide Internet services to customers. Originally there was a clear distinction between data and voice services, but the introduction of technologies such as VoIP has blurred this distinction. As VoIP becomes more common, it is increasingly difficult to distinguish between voice and data traffic. At the level of international gateways, this has created what is referred to as a "gray market," in which companies with international connections ostensibly used for data traffic carry international voice traffic and then pass it on to local networks. Such operations are difficult to detect and regulate, and the resulting intense competition for international voice services has led to a further reduction in prices (Cohen and Southwood 2004).

VoIP is also being used to provide voice services directly to customers. Companies with licenses to provide Internet services, such as iBurst in South Africa, have been using them to provide voice services to end users through VoIP software packages. In the past, the user interface for VoIP services was less convenient than traditional voice telephony because it has required connecting via a computer. VoIP services have therefore tended to focus on high-value voice services such as international calls. As broadband becomes more common, however, VoIP packages are becoming more user friendly, and companies are enabling more traditional telephone handsets to use broadband. This is broadening the impact of VoIP on traditional voice-based businesses.

The legal status of VoIP varies across the region. Some countries ban it, others allow it, and in others, its legality is open to question. For example, VoIP may be legal only for licensed telecommunications operators, but its use is still tolerated for unlicensed users (appendix table A2.2). According to the International Telecommunications Union (ITU 2008, p. 28): "While just over half of countries allow VoIP in some form or another, its legal status is vague in three quarters of the countries. Restrictions against VoIP constrain consumers from making lower priced calls and delay Africa's transition to next-generation networks." This regulatory confusion stems mainly from the difficulty in detecting VoIP traffic and prevents ISPs from offering this service to their customers. Regardless, VoIP operators charge lower prices, which has benefited users of ICT services, particularly in the international call services market.

A recent trend observed particularly in West and Central Africa is the tighter regulation of incoming international voice traffic. In a liberalized market, operators of international gateway facilities compete with one another to carry incoming international traffic into the country. This competition pushes down prices for international termination. Several countries in Africa have signed agreements with companies that monitor incoming traffic through international gateways and impose a fixed termination charge. Typically under these arrangements, the increased revenues are shared between the operator of the gateway management company and the government. Such arrangements have the effect of reversing the policy of liberalization that has been so successful in reducing prices and increasing traffic volumes (Balancing Act 2007). They may also reduce government revenues in the long run as they reduce traffic volumes.

Competition in Internet Provision

Several different business and network elements need to be combined to provide Internet services to customers (figure 3.3). Competition in

Figure 3.3 Broadband Internet Value Chain

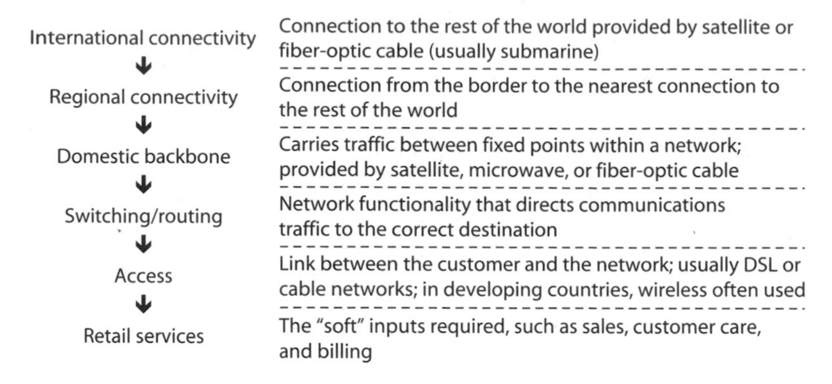

Source: Adapted from Williams 2010.
Note: DSL = digital service line.

the Internet market therefore depends on competition in each of these elements. The most important of these are the access and retail services market, the domestic backbone market, and the international connectivity market.

Internet services. Some degree of competition is found in the provision of Internet services in almost all African countries. The entry costs into this part of the market are fairly low; over 40 countries have issued multiple ISP licenses, and a dozen have issued more than 10 licenses (table 3.2). Some countries have gone further in relaxing their licensing regimes for ISPs by establishing a relatively low license fee, or no fee at all.

Despite this high degree of competition at the retail end of the ISP market, ISPs often face constraints in the way that they operate, either because of the regulatory framework or as a result of the level of competition in other segments of the market. The Internet services market is changing rapidly, however, as mobile operators move into the data services market, primarily through upgrades to their voice networks to provide third generation (3G) services. The customer response to this demand has been impressive. By 2010, South Africa had 5.8 million wireless mobile broadband connections, Nigeria had nearly 10.0 million (according to Wireless Intelligence, a global database of mobile market information), and, in Kenya, Safaricom is reported to have gained 3.4 million 3G subscribers in the two years following its launch in 2008 (Business Monitor International 2010). Mobile operators, when they enter the data services

Table 3.2 Number of ISPs in Sub-Saharan African Countries, 2009

Number of ISPs	*Countries*
1	Comoros, Equatorial Guinea, Ethiopia
2–10	Angola, Benin, Burkina Faso, Burundi, Cape Verde, Central African Republic, Chad, the Democratic Republic of Congo, Côte d'Ivoire, Eritrea, Gabon, The Gambia, Lesotho, Madagascar, Malawi, Mali, Mauritania, Mozambique, Namibia, Rwanda, São Tomé and Príncipe, Senegal, Seychelles, Swaziland, Togo, Uganda, Zimbabwe
11 or more	Botswana, Cameroon, Republic of Congo, Ghana, Kenya, Mauritius, Niger, Nigeria, South Africa, Sudan, Tanzania, Zambia

Sources: Operators, ISPs, and regulatory authorities.
Note: Data are not strictly comparable because some countries report the number of licensed ISPs regardless of whether they are in operation or not. Countries are not shown if data on the number of ISPs could not be obtained. ISPs = Internet service providers.

market, bring with them extensive network coverage, a wide customer base, and significant financial resources. Competition in Internet service provision is therefore likely to intensify as mobile operators increasingly focus on data as a source of revenue.

Domestic backbone market. Chapter 2 discussed the rapid growth of domestic fiber-optic backbone networks and showed how extensive these networks are in some African countries. But how much competition is found in this market segment?

The first way of addressing this question is to consider the patterns of ownership of these networks. As with all types of telecommunications infrastructure, the state-owned telecommunications operators were historically the main or often the only investor in fiber-optic networks in the region. Following the wave of market liberalization and privatization, ownership of fiber-optic backbone networks has become more diverse (figure 3.4).

The private sector currently owns about 59 percent of the total operational fiber-optic network infrastructure in Sub-Saharan Africa; state-owned enterprises (SOEs, which include both telecommunications and electricity companies) own the rest. The picture is slightly different for networks under construction. The private sector is responsible for about 51 percent of these, indicating that the government still continues to play a significant role in the development of fiber-optic infrastructure.

State-owned telecommunications operators are the traditional means of government investment in telecommunications infrastructure; such companies own 72 percent of the publicly owned operational fiber–optic

Figure 3.4 Ownership of Fiber–Optic Backbone Networks

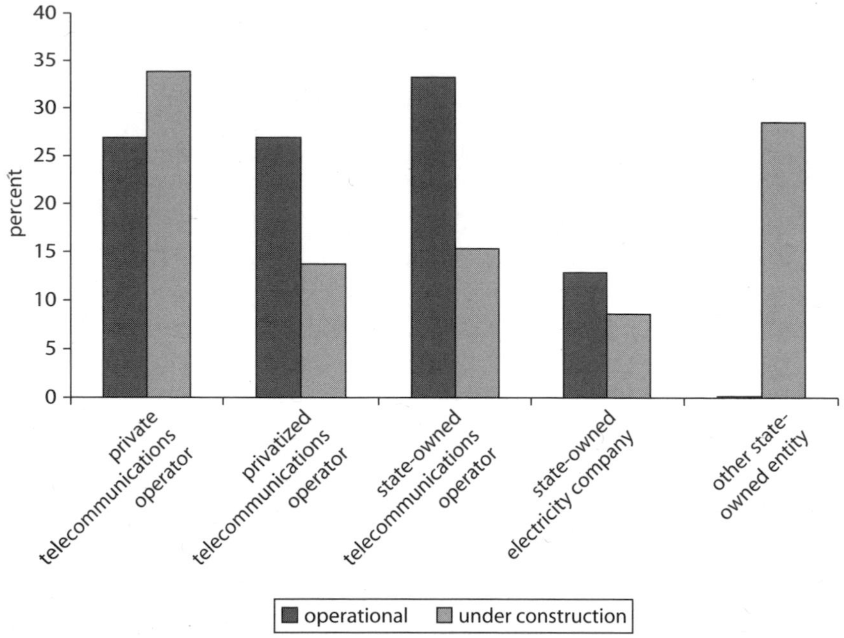

Sources: Hamilton 2010; World Bank staff analysis.

networks, with the remainder owned by state-owned electricity companies. The pattern of new public investment in fiber-optic infrastructure is, however, quite different. Of the total publicly funded fiber-optic infrastructure currently under construction, 54 percent is being implemented directly by governments (as in Kenya, Rwanda, Tanzania, and Uganda); 29 percent via the conventional vehicle—the state-owned telecommunications operators (as in Ethiopia, Mozambique, and Zambia); and the rest by state-owned electricity companies. Despite being newcomers to the telecommunications market in Africa, these electricity companies are having a significant impact: Ghana, Kenya, Malawi, Tanzania, and Zambia are all examples of countries in which electricity companies have upgraded their networks to include fiber-optic networks and used them to provide backbone telecommunications services. They are usually wholesale-only businesses, which means that they are able to provide backbone services to many downstream operators. They also have technical advantages because their networks are typically more secure than networks buried underground.

Across Africa, governments have been disengaging from the sector and relying on private investment and competition. At the same time, they have begun to invest more public resources in fiber-optic infrastructure. Does this mark a generalized shift in policy toward competition in the sector? At this stage, the answer appears to be no. Even where governments have used public resources to invest in fiber-optic infrastructure, the process of sector reform toward private, competitive markets has usually continued. In Zambia, for example, the government has implemented a national fiber-optic backbone network project via the incumbent operator, Zamtel, while at the same time undertaking the privatization of the company. In East Africa, public investment in backbone networks does not seem to have been associated with a general reversal in the liberalization process. In some cases, public investment can actually promote competition. In Rwanda, for example, the government is constructing a national network that incorporates additional ducting that will be made available to operators to lay their own fibers. There are also a few examples of countries where fiber-optic backbone networks owned by different SOEs compete with one another. In Zambia, the electricity parastatal Zambia Electricity Supply Corporation Ltd. (ZESCO) is competing against Zamtel to provide fiber-based backbone services.[12] In South Africa, Sentech and Infraco are both government-owned backbone operators in competition with each other and with Telkom SA, which is partly owned by the government.

Although an overall shift in sector policy resulting from public investment in fiber-optic networks does not seem to be taking place, a long-term risk exists that future policy and regulatory decisions will be influenced by a desire to protect these investments. This has often been the case, evident in both implicit and explicit regulatory protection of state-owned incumbent operators (for example, the nonpayment by incumbent operators of interconnection debts to private operators). Examples have also been seen of it happening specifically with backbone networks. In Botswana, for example, before 2005, mobile operators were required to use the incumbent operator's backbone network where it was available, despite operator complaints of the high cost and poor quality of service (Ovum 2005). In Burkina Faso, the Telecommunications Act of 1998 allowed mobile and other network operators to develop their own backbone networks but prevented them from selling backbone services to one another or to third parties. This reduced the incentives for operators to invest in backbone networks, resulting in a focus by investors on lower-capacity wireless networks (Williams 2010). These regulatory restrictions have since been lifted in Burkina Faso, but similar restrictions remain in place in Mozambique. A more recent example is that of Tanzania, where the government has invested public funds in the construction of a national fiber-optic backbone network, the management of which has been given to Tanzania Telecommunications Company Ltd. (TTCL), the state-owned fixed-line operator. Other operators wishing to develop their own fiber-optic backbone networks have found it difficult to obtain rights of way along the nation's road infrastructure, despite an overall sector policy supporting private ownership and competition. Restrictions such as these, designed to protect public investments in networks, limit investment and constrain the development of competition in this vital segment of the market.

The construction of a government-funded network therefore has the potential to stimulate private sector investment and competition, rather than displace it, but it could also have the opposite effect, by substituting for private investment and skewing the regulatory environment to protect the state's investment.

Where multiple networks are being built, is competition among them—considering their high fixed and sunk costs—economically feasible? Relatively few countries have fully liberalized their fiber-optic backbone market and actively encouraged the development of competition, so the track record of competition across the region is not as well established as in the mobile market. However, the few countries that have led the way in encouraging investment in fiber-optic networks have seen the rapid emergence of infrastructure competition in this market segment. Nigeria has been most successful in this area. Its explicit efforts to attract backbone operators through the issuing of licenses specifically for backbone networks (often referred to as "carrier licenses") have spurred investment in fiber-optic networks, which compete directly with each other. Here mobile operators have also invested in upgrading their networks to fiber to be able to carry the high volumes of traffic that their customers generate (MTN 2006) (figure 3.5). Other countries that have fully liberalized their infrastructure markets, such as Kenya and Rwanda, have seen similar patterns of network development.

Infrastructure competition is also driving cooperation in network deployment. This cooperation comes in several different forms. In Nigeria, for example, competing operators have entered into agreements to share facilities when doing so is mutually beneficial. Two main types of such sharing arrangements are found: (1) geographical swapping, in which two

Figure 3.5 Fiber-Optic Backbone Networks in Nigeria, 2009

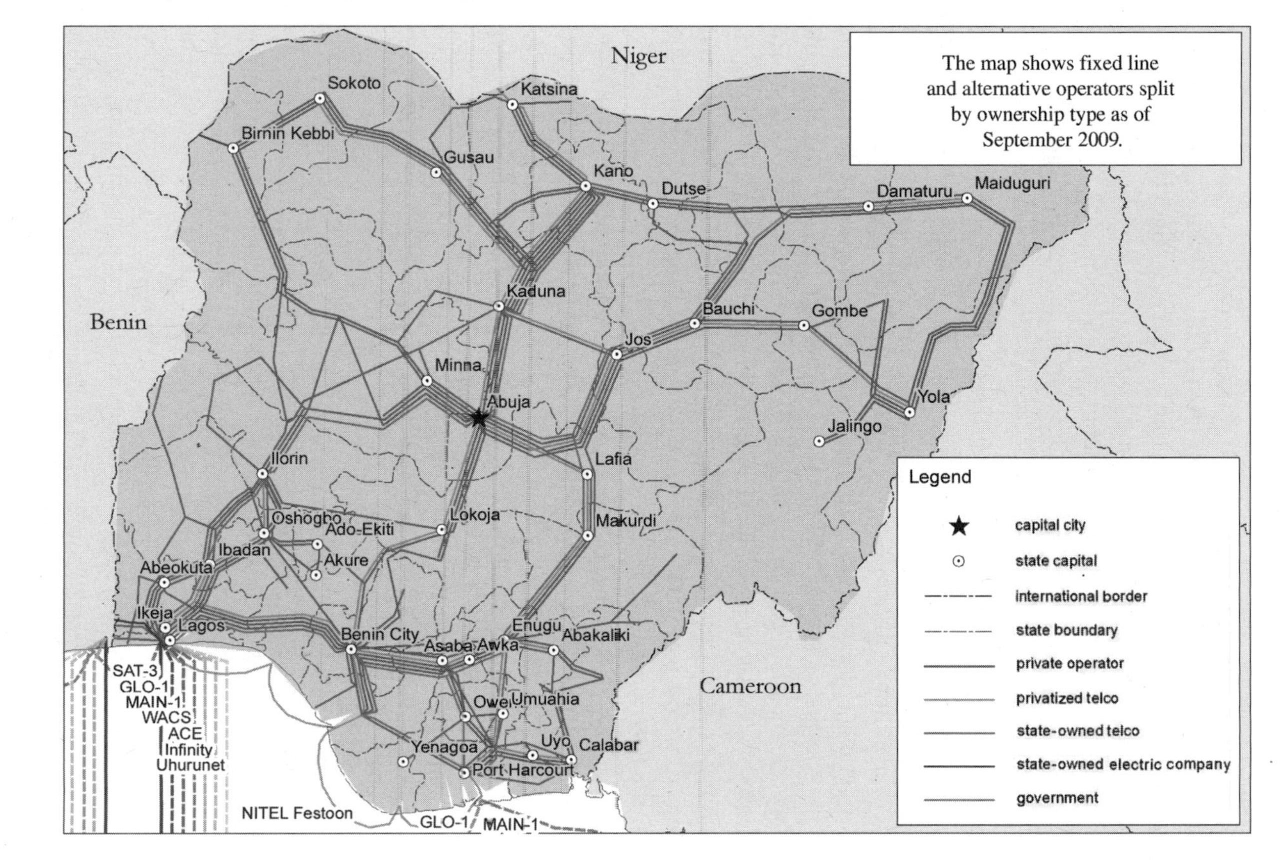

Source: Hamilton 2010. © Hamilton Research Ltd.

competing operators build networks in different parts of the country and then agree to swap capacity on their networks; and (2) route swapping, in which operators with competing networks along the same routes swap capacity or fibers on those routes. In other countries, examples are seen of joint network construction projects in which competing operators plan and execute fiber-optic network rollout projects together. In South Africa and Tanzania, for example, competing operators are collaborating on joint national fiber-optic backbone projects, and in Zambia, MTN and Zain are collaborating on a joint fiber-optic network in the capital, Lusaka.

These cooperative arrangements can increase the geographical coverage of networks, lower costs, and improve service quality by providing operators alternate routes for traffic in the event of technical problems on their own networks.

Unlike mobile networks in the region that cover large segments of the population, competition among fiber-optic network operators is concentrated on the most profitable routes—those that connect major population centers and those that connect to the landing points of submarine fiber-optic cables in countries that have them. In Sudan, for example, the landing point for fiber-optic cables is Port Sudan, which is not the main population center, but competition is seen between backbone networks on the Khartoum–Port Sudan route (figure 3.6).

The high up-front investment costs involved in fiber-optic backbone networks make them commercially viable only in areas where traffic levels are high or are likely to increase (Milad and Ramarao 2006). Fiber-optic networks connecting small towns and villages therefore seem unlikely without some form of government intervention (Williams 2010). African governments are experimenting with different types of such interventions. In Kenya, the government is supporting network investment in areas off the main trunk routes connecting Mombasa to Nairobi and from Nairobi to the borders of Tanzania and Uganda, while in Rwanda, the government's construction of a national backbone network, which included additional ducting to facilitate private operators in building their networks, should boost competition. In Ghana, the government invested in extending the fiber-optic network controlled by the state-owned electricity transmission utility (the Volta River Authority [VRA]) and then included the communications assets in the privatization of the incumbent telecommunications operator, Ghana Telecom. The challenge facing all these schemes is how to use public resources to boost overall investment in backbone networks without displacing private investment or adversely affecting competition.

Figure 3.6 Fiber-Optic Backbone Networks in Sudan, 2011

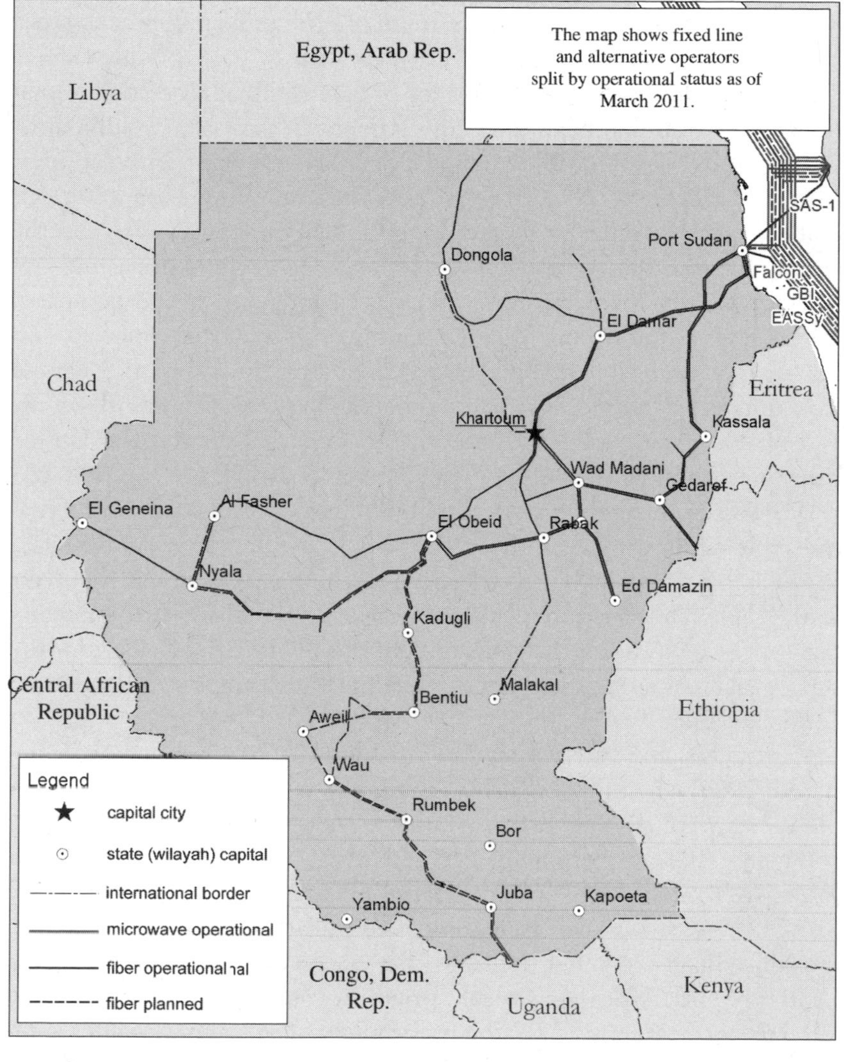

Source: Hamilton 2010. © Hamilton Research Ltd.

International connectivity. The liberalization of the international facilities markets has allowed mobile operators and ISPs to invest in networks that can carry traffic in and out of countries. The resulting competition has lowered costs and wholesale prices, which in turn has reduced the retail prices of international calls and Internet services. However, traditionally across most of the continent, this international connectivity has been via satellite link. The supply of satellite bandwidth to Africa has been controlled by a few satellite operators,[13] and satellite wholesale bandwidth prices have remained relatively high.

The only alternative to satellites for obtaining direct access to international bandwidth has been the South Atlantic 3/West Africa Submarine Cable (SAT-3/WASC) cable running along the west coast. This cable has a design capacity of 120 gigabits per second, and its launch in 2002 dramatically increased the amount of international capacity available to the region. Despite this, international wholesale bandwidth prices did not fall significantly following its launch. Even today, after the launch of other cables in the region, prices remain high. The average price of leasing an international E1[14] link between South Africa and the United Kingdom, for example, was approximately $13,000 per month at the end of 2009, compared with $4,000 per month between the United Kingdom and India (Mumbai) (TeleGeography 2010).

The reason for these historically high prices of bandwidth on submarine fibre–optic cables has been the lack of competition. The SAT-3/WASC cable is owned by a consortium of companies that control access to the cable. The structure of the consortium has allowed it to maintain high prices and prevent the development of a competitive market for the supply of submarine fiber-optic-cable–based international bandwidth (Akoh 2008).

Customers in South Africa have suffered from a similar problem. Before 2009, when the SEACOM cable was launched, although multiple cables were landing in the country, they were all controlled by one company—Telkom SA. This ensured that international bandwidth prices remained high by international standards. The launch of SEACOM—access to which is controlled by Neotel, a competitor—has introduced some competition into the market for international bandwidth in South Africa. As a result, prices in South Africa have begun to fall. Further competition will be introduced when the new cables—the West Africa Cable System (WACS) and Africa Coast to Europe (ACE)—are launched.

In contrast to the situation along the west coast of Africa, in regions of the world where competition between submarine fiber-optic cables has developed, wholesale prices have fallen dramatically. The average price of a link between Miami in the United States and São Paulo in Brazil, for example, fell by 70 percent between 2002 and 2009 (TeleGeography 2010). A similar situation is beginning to emerge on the east coast of Africa, following the introduction of three new cables with different owners in 2009 and 2010. Within a year of the launch of the first of these cables, SEACOM, international bandwidth prices had already begun to drop. Wholesale bandwidth prices on submarine cables in East Africa are not published; therefore, it is difficult to obtain definitive benchmark prices, particularly in regions where the market is new. Discussions with operators in the East Africa region indicate that prices had fallen by 50 percent in the first year since the launch of SEACOM, and even further for purchases of large capacity units. Market observers expect prices to fall further as competition intensifies.

For the majority of African countries, the high levels of current and planned future investment in submarine fiber-optic cables will meet the capacity requirements of the market for the foreseeable future. Customers will benefit from this, however, only if effective competition is established. The ownership of the new cable infrastructure in Africa is relatively diverse (see chapter 2), which means that one of the factors that allowed operators to retain high international prices—exclusive control over access to the cables has been removed. Such competition is already beginning to have an impact. In South Africa, for example, the wholesale price of an STM-1[15] link between Johannesburg and London cost nearly $1 million per month in 2005 but had fallen to $145,000 per month by the end of 2009. A similar drop in prices could potentially take place across the region with the emergence of a competitive submarine fiber-optic cable market.

Despite the positive outlook for users of the submarine fiber-optic cable infrastructure in Africa, some challenges remain. The first of these, affecting many countries, relates to regulation of the landing facilities. If competition across submarine cables is to be effective, users need to be able to access the cables easily and at low cost. Cable-landing facilities are an essential part of this equation because they are points of access to each cable. In countries where there are multiple landing points, one for each cable, for example, there is likely to be competition among the landing stations, which in turn benefits customers. However, if multiple cables land at a single facility and that facility is owned by a single party, it creates the conditions for that party to exploit its position. It also leaves the cables vulnerable to any technical problems at the landing point. In such cases, strong regulation is needed to ensure that the owner gives all users equal, cost-based access, but international experience indicates that this is difficult to implement. The government of Singapore, for example, spent several years trying to impose regulated access conditions on the incumbent operator Singtel's cable-landing facilities, until finally including access to them on regulated terms in Singtel's Reference Interconnection Offer (IDA 2004). India underwent a similar process to force down prices of international bandwidth (referred to as international private lease circuits [IPLCs]) (Esselaar, Gillwald, and Sutherland 2007). The dependence on strong regulatory measures can be reduced in Africa by ensuring that each country has multiple, competing landing stations. In addition, joint ownership of the landing stations by interested parties may give scope for some self-regulation. Finally, strong regulatory rules that enforce cost-based open access to landing facilities should be introduced to ensure that the benefits of the additional capacity provided by submarine cables are felt by customers.

Competition is also developing in the provision of international bandwidth to countries that do not have direct access to submarine cables. Carrier networks are developing cross-border infrastructure as part of their regional development strategies. In West Africa, for example, Suburban Telecom is a Nigerian network operator with a national fiber-transmission backbone. It has also built a cross-border fiber-optic link to neighboring Benin, and, in June 2009, the company announced that it had completed the deployment of a regional fiber-optic network running from Nigeria to Ghana via Benin and Togo. Phase 3 Telecom is another Nigeria-based fiber-optic backbone network operator and has plans to extend its network into Benin, Côte d'Ivoire, Ghana, Senegal, and Togo using the high-voltage power transmission lines in these states. Similar developments are taking place in the east and south as backbone networks link up across borders.

This development of competition in the regional network infrastructure is being stimulated by the arrival of more submarine fiber-optic cables. The high prices charged by SAT-3 cable operators drive terrestrial networks to build across borders so that they could play one off against another. Suburban Telecom's connection from Nigeria to Benin is an example of this. One of the operator's objectives was to provide Nigerian customers with access to Benin's SAT-3 submarine-cable–landing station, thereby bypassing exclusive access by Nigerian Telecommunications Ltd. (NITEL) to the cable in Nigeria. On the east coast, the lower prices for international bandwidth brought about by the arrival of the submarine cables have strengthened the commercial case for supplying this bandwidth to landlocked countries in the region. This has stimulated the development of fiber backbone networks along the Mombasa-Nairobi-Kampala-Kigali route, which are currently under development by at least two operators (Altech Stream East Africa and LapGreen Networks).

Finally, the presence of telecommunications operators such as Etisalat, LapGreen, Maroc Telecom, MTN, and Zain, with operations in multiple

African countries, is also driving the development of regional backbone networks. By building network infrastructure across borders to link operations in neighboring countries, these regional operators are evolving from being groupings of stand-alone businesses to regional operators that compete to carry long-haul traffic between countries. MTN, LapGreen, and Altech Stream East Africa are all examples of this in East Africa.

Despite these promising signs for competition in cross-border terrestrial infrastructure, there remain many major gaps in the continent's regional backbone networks. Accessing submarine-cable-landing points remains a major challenge for many landlocked countries in particular.

Private Sector Participation

The main driver of telecommunications network expansion in Africa has been investment, most of which has originated from the private sector. With the exception of the Comoros, Eritrea, and Ethiopia, all of the countries in Sub-Saharan Africa had allowed foreign investment in their telecommunications sector as of the end of 2009, and most allowed foreigners to own at least 51 percent of a given company. Various countries have gone further by granting foreign investors complete ownership of subsidiaries.

Private investors have entered the market in two ways: by (1) purchasing state-owned operators through privatization and (2) investing in greenfield projects following the acquisition of a license.

Privatization

The privatization of state-owned incumbent operators has been a steady trend in Africa since the mid-1990s and has provided an entry point for private investors to participate in the telecommunications sector. Governments raised $4.6 billion through telecommunications privatization transactions between 1995 and 2008.[16]

Most privatization has been done through an equity purchase by a major operator that then takes management control, but notable exceptions to this are seen. In Sudan, for example, the government has released its holdings on the Khartoum and regional stock markets in several sales since 1993.[17] In Kenya in 2008, the government sold a 25 percent stake in Safaricom, the leading mobile operator, through an initial public offering on the Nairobi Stock Exchange. The government retained a 35 percent stake in the company.

In the late 1990s and 2000s, investors from developed countries were significant participants in purchases of incumbent operators through privatization. Between 1995 and 2000, for example, state-owned operators in Cape Verde, Côte d'Ivoire, Senegal, and South Africa were partly divested to strategic investors from France, Portugal, and the United States. Following this initial period, however, investors from developed countries largely withdrew from the privatization of telecommunications in Africa. Recent privatizations have involved public offerings (in South Africa and Sudan), sales to developing-country investors (in the sales of incumbent operators in Burkina Faso, Gabon, and Mauritania to Maroc Telecom), and sales to domestic investors (in the case of Malawi Telecommunications Limited). Two recent high-profile transactions may signal the return of developed-country investors to the telecommunications market in Africa: the sale of 51 percent of Telkom Kenya to France Telecom in December 2007 and the sale of 70 percent of Ghana Telecom to Vodafone (United Kingdom) in August 2008.

Privatization has not always proceeded smoothly. For example, the government of Nigeria has made repeated unsuccessful attempts to privatize NITEL. A tender for 40 percent of NITEL was issued in 2001 with the winner announced in 2002, but neither the winning bidder nor the second-place bidder was able to pay the final bid price. The government later announced its intention to sell the company by 2005. In 2006, Transcorp purchased 51 percent of NITEL after failing to pay the full amount for a 75 percent stake. In 2009, a new tender for 75 percent of NITEL was announced with Transcorp, apparently being required to relinquish some of its stake to a new owner. The winner was announced in February 2010, but the transaction was then further delayed pending a review (Osuagwu 2010). In Guinea and Rwanda, the government privatized the state-owned operator but subsequently renationalized it by repurchasing shares. In the case of Rwandatel, the government subsequently resold the operator in 2007. In Guinea, the government repurchased its incumbent operator Sotelgui from Telekom Malaysia in 2008, which had bought it in 1996.

In a few countries, governments have opted to forgo privatization in favor of other forms of private participation. The governments of Botswana, Ghana, Nigeria, and Tanzania all signed management contracts with private companies to run their incumbent operators while retaining state ownership. The most recent example of such an arrangement in Africa is the awarding of a management contract for Ethiopia Telecom to Sofrecom in 2010.

Management contracts for state-owned operators in Africa have often not been successful, frequently ending in premature cancellation. Many reasons for these failures can be identified, but they usually involve accusations of political interference in business operations, on the one hand, and failure to invest, on the other.[18]

The privatization of state-owned telecommunications companies has long-term implications for the sector. The immediate effect is to raise government revenue upon initial sale. A medium-term effect is avoiding the economic cost of state ownership arising from nonpayment of taxes and the low productivity of state operators. Privatized operators typically grow as money is invested in them and become more efficient. This generates economic growth and increases government tax revenues, but the long-term benefit of privatization is felt by the sector as a whole. Once an operator has been sold to private investors, little incentive is seen for the government to bias regulatory or tax rules in its favor. This creates a more competitive market, which in turn is more efficient and more productive and thus benefits the wider economy.

Greenfield Investments

The majority of private investment in the telecommunications sector has been through greenfield investments. These result from the acquisition of a license—usually some kind of mobile telecommunications license or a data services/ISP license—either procured directly from the government or through the acquisition of an existing licensee.

A wide range of investor types can be identified; the size of their investments also varies, from the usually small investments made in ISPs to the very large ones made in mobile operators (see chapter 4). A significant recent trend in the mobile segment of the telecommunications market in Sub-Saharan Africa has been the emergence of multinational mobile operators. These pan-African investors account for over 80 percent of mobile subscribers in the region. The top eight pan-African investors accounted for about three-quarters of mobile subscribers in the region in 2009 (table 3.3).

A summary of the commercial interests of these multinational operators in Africa appears in appendix table A2.7. The emergence of large regional operators has benefited telecommunications services in the region. Regional operators have experience working in African markets and therefore can easily access technical and operational know-how. Economies of scale in the purchasing of network equipment also help to reduce costs. A related trend is the development of cross-border network connections, discussed above, which are facilitating the development of regional backbone networks. The benefits of these regional operators were highlighted in 2005 by the Zambia Competition Commission (ZCC), which approved the sale of local mobile operators to Mobile

Table 3.3 Top Multinational Mobile Investors in Sub-Saharan Africa, by Number of Subscribers, 2009

Operating group (country)	*Brand*	*Subscribers (million)*	*Number of countries*
MTN (South Africa)	MTN	82.8	16
MTC (Kuwait)[a]	Zain[b]	50.6	15
Vodacom (South Africa/United Kingdom)[b]	Vodacom	48.5	7
France Telecom	Orange	21.2	14
Millicom (Luxembourg)	Tigo	12.2	8
Portugal Telecom	[c]	8.6	5
Etisalat (United Arab Emirates)	Moov	4.3	7
Maroc Telecom (Morocco)	[c]	4.2	4

Source: AICD, adapted from company and regulator reports.
Note: Except where noted below, subscribers are the total in each operation and not proportionate to the investors' shareholding. MTC = Mobile Telecommunications Co.
a. MTC's African holdings except for Sudan were sold to Bharti of India in 2010.
b. Subscribers for Kenya are proportionate because there is no controlling interest.
c. Local brand names are used.

Telecommunications Co. (MTC) and MTN.[19] As noted by the ZCC, strategic mobile investors offer many benefits, including access to capital, knowledge and technical ability, and purchasing strategies that lower costs and increase roaming possibilities, allowing mobile subscribers to use their phones in a larger number of countries.

Regulation

Developing competition in the ICT market requires more than just issuing licenses. It also requires establishment of a regulatory authority and the development of rules and regulations to govern the sector. These regulations evolve as competition develops and sector priorities change. The development of regulatory institutions and the associated regulatory framework are therefore key aspects of market liberalization and sector reform.

Regulatory Institutions

The first NRA in Africa was created in 1992 in Nigeria, and, according to the ITU, by 2009, 41 countries in the region had NRAs (appendix 2A.8). These institutions have usually been set up as quasi-independent of government with the intention of letting them make technical decisions without undue political interference.[20] In practice, the degree of independence varies significantly across countries.

This independence depends on many factors. Particularly important are the ways in which the institution is financed and the senior staff selected as well as terms of employment. Most NRAs in the region are financed through levies on telecommunications operators (such as fines, penalties, license and spectrum fees, and special taxes on revenues). This generally ensures a reliable source of income, given the rapid development of telecommunications networks in the region. At the same time, the mechanism used to determine the NRA's budget affects its independence. In some cases, the NRA budgets have to be approved by the administration, and usually by the legislature as well. The NRA collects license fees and levies, which are then generally remitted to the government. The NRA can either retain the amount required to fund its budget before remitting the rest, or pass the full amount on to the government, which then returns the requisite funds to the NRA. In a few cases, NRAs are fully funded directly by the government budget, and any funds (such as license fees) that they collect are remitted to the government in their entirety (figure 3.7).

The more direct control a government has over the budget of an NRA, the more vulnerable it is to political interference in its decisions. The terms of employment for senior staff also affect the independence of a regulatory authority. In some countries, the sector minister appoints the chief executive or the board members of an NRA, which leaves these staff vulnerable

Figure 3.7 Financing Structure of Regulatory Authority for Select Countries in Sub-Saharan Africa, 2007

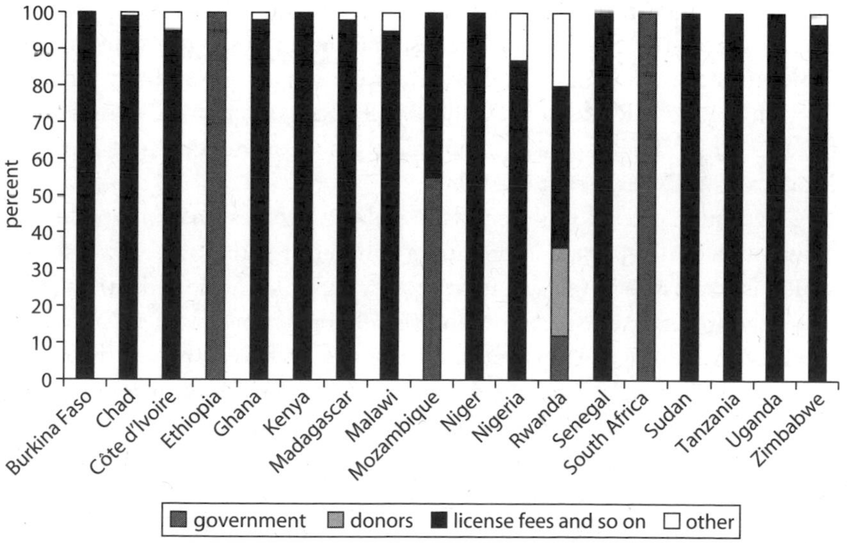

Source: Ampah and others 2009.

to political influence, especially if the same government ministers can terminate their employment. Ideally, the senior executives of the NRA should be appointed by senior levels of government such as the president, prime minister, or the cabinet and should be protected by fixed-term appointments. Among countries reporting this information, 43 percent of NRA heads are appointed by the head of state and 22 percent each by the sector minister or council of ministers. Parliament appoints the NRA head in the remaining 13 cases (Ampah and others 2009).

Making NRAs responsible for more than one sector can also increase their independence from direct political influence over any one sector—whether from the government or industry players. By increasing the scope of responsibilities, however, the NRAs risk not being able to focus adequately on the regulation of specific sectors. In practice, only a few countries have established multisector regulators—Gambia, Mauritania, Niger, and Rwanda have regulators responsible for other utilities, such as electricity and transport as well as telecommunications.

Despite this focus on the telecommunications sector in most regulatory institutions in Africa, their scope of responsibility is evolving. As telecommunications technology and businesses have evolved to include media content in the services that they deliver, regulatory institutions have also evolved to include nontraditional services such as the media and postal services in their scope of responsibility (Shannon 2006; Singh and Raja 2010). Several Sub-Saharan African countries have moved in this direction, for example, the Independent Communications Authority of South Africa (ICASA), which was established in 2000 from the merger of the telecommunications and broadcasting regulators, and the Tanzania Communications Regulatory Authority (TCRA), which was formed from the merger of the Tanzania Communications Commission and Tanzania Broadcasting Commission in 2003.

The significance of this trend is less about institutional independence and more about focusing on providing coherent regulation across a sector, which is itself integrating what were previously separate industries (that is, telecommunications, information technology, broadcasting, and postal services). Matching the institutional arrangements to trends in the sector can also, however, create difficulties for an NRA. Telecommunications regulation has traditionally focused on technical, economic, and legal aspects. Broadcasting and media regulation, however, is typically more politically sensitive and can involve regulation of content.

Accountability and transparency are essential to an effective regulatory authority. African telecommunications sector regulators are making progress on this front, particularly when compared with other infrastructure sectors. A review of regulators in southern Africa that assessed the extent and quality of NRA websites showed that over half publish information online about the sector, thereby increasing transparency. The report recommended that regulators provide a comprehensive, up-to-date, and easily accessible repository of reliable regulatory information for potential investors and stakeholders. Ideally, they can use websites to clarify rules, regulations, and regulatory processes for users and stakeholders; provide easy access to forms for regulatory processes; assist in communicating directly to required departments; and facilitate stakeholder compliance and reporting.

Set against this benchmark, the availability and quality of information on African regulators' websites is inconsistent. The review concluded that: "The type of information provided across the African sites also raises issues of concern. While there is a remarkable improvement from the last review done, information still remains largely factual with very little effort made to explain and allow the reader to digest the information provided. … Disappointing across all the sites was a lack of effort made to analyse the statistical information that was laid out on the websites. This type of information would be very informative to many stakeholders and in addition provides information for researchers and journalists internationally on the development growth of a country. In addition, except for information regarding licensing procedures, many websites lacked the information usually sought by businesses and investors … Presenting information that was useful to consumers is another category that was also neglected by the majority of the African NRA's … In addition, content on quality of service (QoS) parameters and health and environment issues were covered by fewer than three of the 30 NRSa reviewed…" (Kerrets-Makau 2009, p. 58).

This inconsistency is also found in the effectiveness of regulatory oversight and governance across the region. Figure 3.8 provides a scorecard for African countries grading the performance of telecommunications regulatory authorities as well as other sector-related issues across four dimensions (Vagliasindi and Nellis 2010). The reform dimension includes efforts to liberalize the sector, such as corporatization of the incumbent, introduction of private participation, and revision of telecommunications laws. Regulation includes characteristics of the NRA, such as autonomy, financing, and the process of appointing senior staff, and the NRA's policies on price regulation, interconnection, and competition. Finally, the governance dimension grades the incumbent operator's autonomy, transparency, and market discipline. Each item within each dimension receives a score of 0 or 1. The binary scores are then averaged to obtain the overall dimension

Figure 3.8 Scorecard for Oversight of the Telecommunications Sector in Sub-Saharan Africa, 2007

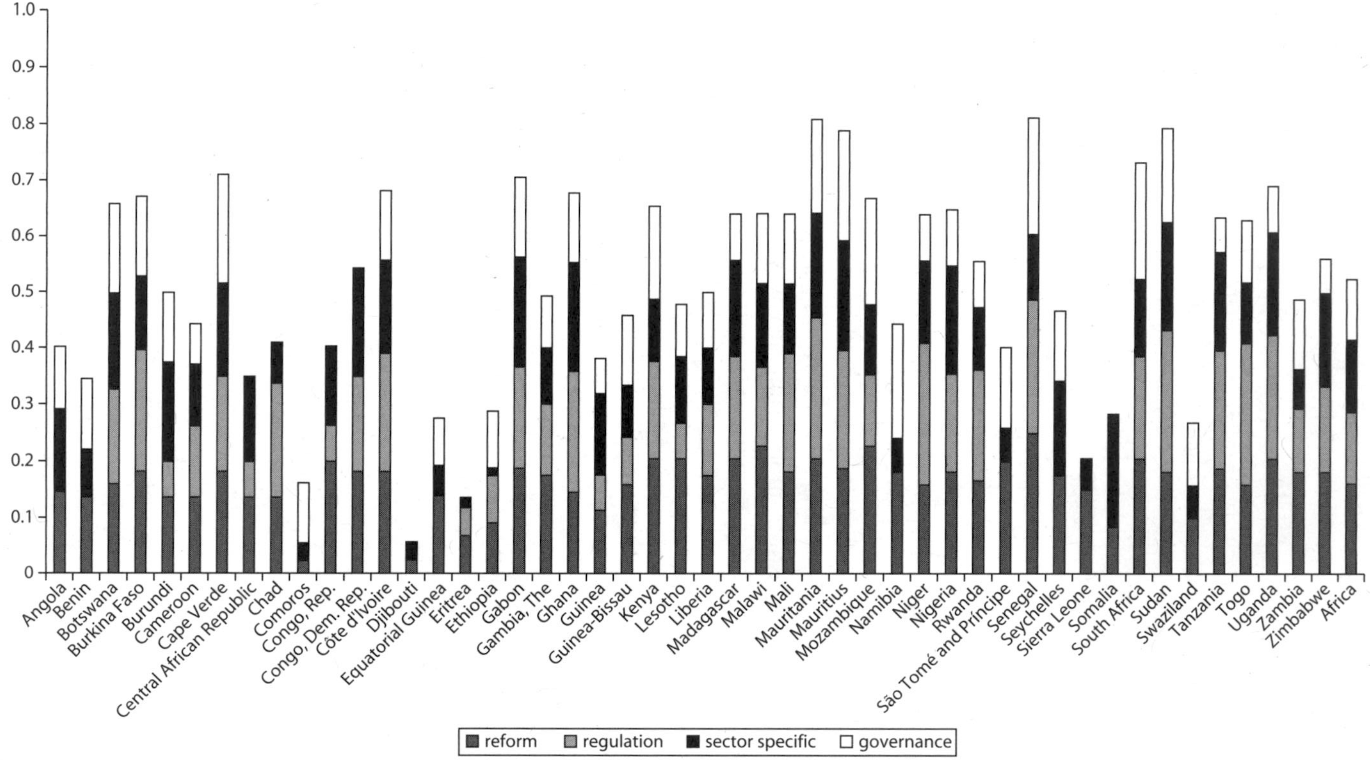

Source: Ampah and others 2009.

score, which is then weighted by 25 percent to get an overall country score. Data were obtained from NRAs and ITU regulatory studies.

It is clear from figure 3.8 that the performance of telecommunications regulatory frameworks varies considerably across Africa. A country's regulatory system requires more than just institutional change for it to be effective. How such institutions are established and managed in large part determines how well they function.

Regulatory Issues

Countries across the region have followed a similar path of sector liberalization, albeit at different speeds. NRAs therefore face many common challenges. Three of the most important ones are interconnection regulations, universal services, and radio-spectrum management.

Operators are required to interconnect to one another by law and as a condition of their license, but the charging arrangements for this interconnection are often left either to commercial negotiations or are subject to regulation.[21] Commercial negotiations usually result in termination rates being set well above cost and far above the rate charged for terminating calls on fixed networks. This raises retail prices and may have an adverse impact on competition (see chapter 2 and the first section of this chapter). These issues have challenged regulatory agencies, which have often lacked the necessary legal and technical tools to resolve them.

MTRs are regulated in most high-income countries where there is an interconnection payment system.[22] Regulators in Africa have also moved in the same direction by imposing cost-oriented regulatory controls on MTRs. One of the first of these was in Botswana, where the regulator resolved a dispute between the two mobile operators by imposing a benchmark rate based on the termination charges of several European mobile operators, arguing that attempts to derive rates based on costs in Botswana would have taken too much time.[23] More recently, Kenya and Tanzania have sought to control mobile interconnection charges by establishing a rate cap. In Tanzania, the NRA made a decision to reduce MTRs gradually on an annual basis (TCRA 2004). In Kenya, the NRA issued a determination on interconnection charges in 2007 and then reduced them further in 2010 (Communications Commission of Kenya 2010). Nigeria's NRA has also intervened several times by establishing MTR targets (Nigerian Communications Commission 2006).

The regulation of MTRs is gradually being brought within a more general competition regulatory framework, following the evolution of regulation seen in Europe since liberalization began. This starts with an assessment of whether an operator is in a "dominant" position in the market before applying appropriate regulatory measures. The Senegalese NRA conducted a market-dominance analysis of the entire telecommunications sector in 2006, which determined that the incumbent provider was dominant with an 89 percent share of overall sector revenues (ARTP 2006). In Niger and Rwanda, leading operators have also been found to be dominant; the leading operator in Niger was required by the NRA to publish an interconnection catalog (Autorité de Régulation Multisectorielle 2005). These developments are in line with a global trend whereby NRAs regulate the MTRs, even in otherwise competitive markets.[24] Figure 3.9 provides a summary of interconnection policies throughout Sub-Saharan Africa. Country-level information on these policies is provided in appendix table A2.9.

The move toward cost-based interconnection rates will benefit consumers in several ways: by lowering the price of calls, by reducing the gap between mobile and fixed interconnection rates (which will boost fixed-line use, making it a more effective competitor to mobile networks, and enhance the prospects for fixed broadband), and by minimizing

Figure 3.9 Interconnection Policies in Sub-Saharan Africa, 2008

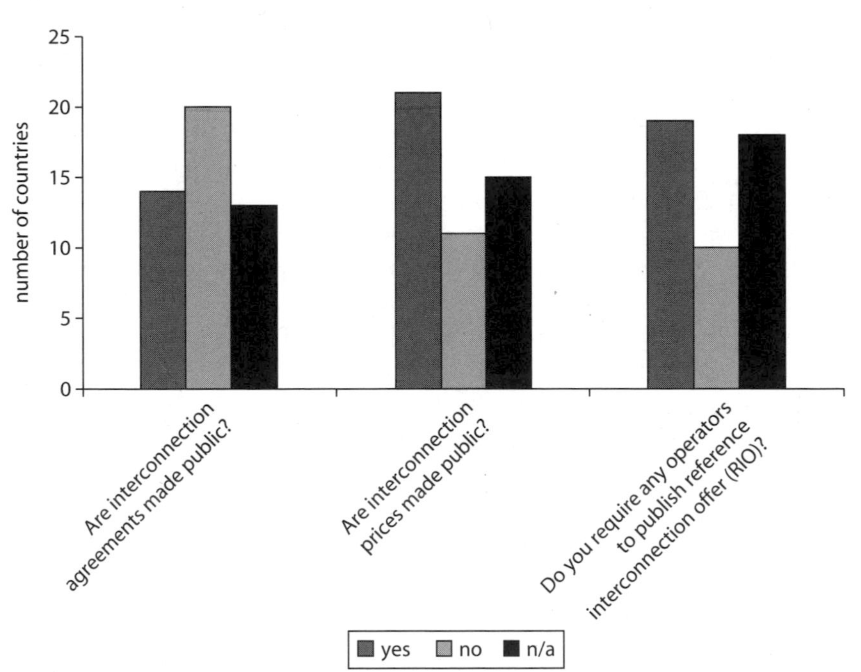

Source: ITU 2009.
Note: n/a = not available.

on-net/off-net price differentials, which may increase the intensity of competition (see discussion of this in chapter 2). A comparison of MTRs throughout Sub-Saharan Africa can be found in figure 3.10.

Internationally, regulatory efforts have pushed down interconnection rates. In India, for example, recent decisions have brought MTRs down to about $0.005 (TRAI 2009). In the European Union, recent moves by the European Commission are likely to push MTRs significantly below current levels (European Commission 2009). This international trend, together with the wide variation in rates seen across Africa, indicates that extensive further regulation of interconnection rates is to be expected.

Another key area of regulatory responsibility is the design and implementation of policies and mechanisms to promote universal access to telecommunications services. In practice, the most successful driver of universal access has been market liberalization, although this is usually not defined as a universal access policy. With the exception of a few small countries, most countries in the region are unlikely to reach 100 percent population coverage based on competition alone (see chapter 5 for a more extensive discussion of this) because a portion of the population lives beyond the commercial reach of telecommunications networks. To promote coverage in areas that are not commercially viable, governments have devised two types of mechanisms: license obligations and universal service funds.

Mobile licenses and privatization agreements often include network-rollout obligations whereby operators have to install a specific number of lines or cover a stated percentage of the population by a certain period. The extent of coverage requirements included in mobile licenses varies. In Africa, as in most other regions of the world, they generally do not require 100 percent population coverage but sometimes include other obligations, such as the provision of pay phones or public calling centers. In South Africa, for example, mobile licensees were required to provide network coverage for 70–80 percent of the population within four years of their launch as well as to meet targets for pay phones and other community telephone facilities (Lewis 2010).

Although network coverage obligations are a simple way to promote universal coverage, they are difficult to modify after licenses have been awarded. Many countries in the region have therefore adopted an alternative mechanism—the universal service fund. Operators are required to contribute a portion of their revenues to the fund, which is then used to subsidize services in areas of the country that would otherwise not be commercially attractive. Disbursement of funds has often been slow, however, and where the funds have been spent, the results have been

Figure 3.10 Mobile Termination Rates in Sub-Saharan African Countries, 2009

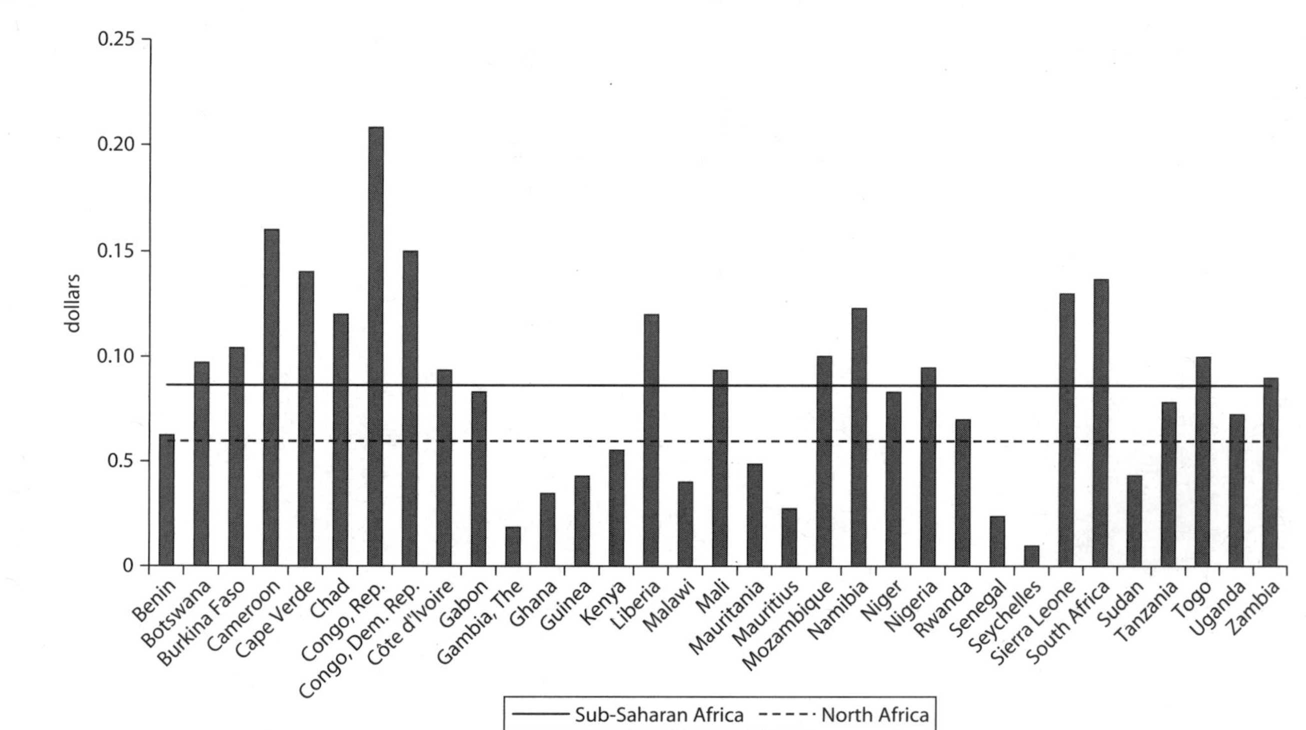

Source: Data from operators and regulators; Ampah and others 2009, updated.

mixed. Uganda is an example of a country where some success has been seen. Each operator in the country contributes 1 percent of its turnover annually to the Rural Communications Development Fund (RCDF), which is used to finance ICT projects outside urban areas. The Uganda Communications Commission administers the program. Between 2003 and 2007, the RCDF financed 76 Internet points-of-presence, 55 Internet cafés, and 3,349 public pay phones in rural areas (UCC 2010). Details on universal service funds in the region are provided in appendix table A2.10.

The private sector has developed a third approach to serving noncommercially viable areas: By establishing shared access points, customers can access ICT services on a call-by-call basis. In Rwanda and Uganda, MTN has launched village pay phone projects, modeled after a successful program in Bangladesh[25] in which rural inhabitants received microfinance to buy mobile phones and then sell airtime to the public. Approximately 9,000 mobile public phones were in use in Rwanda by 2007. In Uganda, MTN had installed 10,000 "Village Phones" by December 2007[26] and was considering similar schemes in other African countries. Vodacom also offers community service telephones: By March 2007, 95,000 had been installed in South Africa, 28,000 in the Democratic Republic of Congo, 4,000 in Lesotho, and 10,000 in Tanzania.

Access to radio spectrum is an essential requirement for new entrants into the mobile market and, increasingly, into the data services market. The demand for this spectrum grows with the number of subscribers and operators, but the supply of radio spectrum is fixed, so it has to be rationed across users. The efficiency with which the spectrum is allocated to users has a major impact on the development of the industry—as well as direct financial implications for governments. Investors are typically willing to pay large amounts of money for some portions of the radio spectrum both at the time of allocation and on a regular basis in the form of spectrum license fees. The allocation and management of radio spectrum among different users is therefore a pressing priority for governments and regulators.

Governments, in line with international recommendations, have typically decided the usage of specific bands of spectrum and then allocated individual slots to new licensees. The price charged for these bands has often been set arbitrarily but, in general, has risen as perceptions of their value have gone up. Since 2000, however, a trend has been seen toward setting prices for spectrum and mobile licenses via auctions and thus allowing the market to determine the prices paid by new entrants.

In this more flexible, market-based system for the allocation of radio spectrum, bands can be bought and sold by operators (Wellenius and Neto

Mobile prices fell in the countries following liberalization, but the price of international calls dropped the most (figure 3.12, panel a).

The result is that, by 2009, countries that had reformed early had lower mobile tariffs and higher fixed-line tariffs than other countries (figure 3.12, panel b).

Market reform also affects the overall size of the industry. In 2000, the total revenue of the telecommunications industry in both groups, relative to GDP, was similar. By 2008, the telecommunications industry in the early-reforming countries generated revenues equivalent to 5.9 percent of

Figure 3.12 Impact of Reform on Tariffs

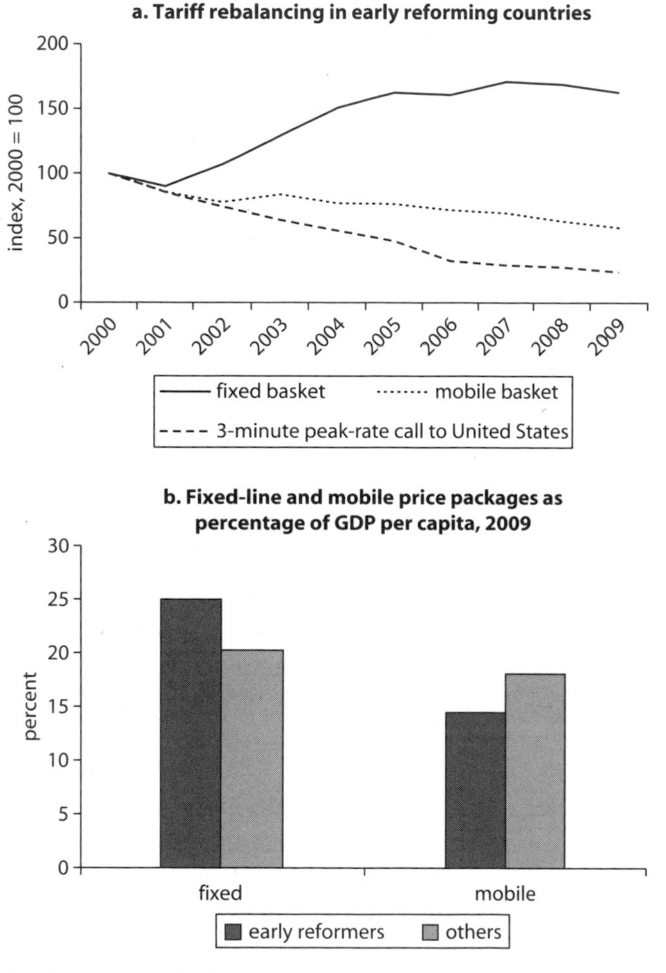

Source: Ampah and others 2009, updated.

GDP, compared with 4.6 percent in the other countries. The industry also grew faster in countries that were early reformers—telecommunications revenues rose by 3.2 percentage points of GDP between 2000 and 2008, compared with 2.0 percentage points in the other countries (figure 3.13).

The numbers of operators and their concentration in the market also have important effects on access and pricing. As the number of mobile operators increases and competition for customers intensifies, the market grows more rapidly. The year-to-year gains, measured in terms of new subscribers, increase significantly after a second operator enters the market and then continue to increase as more operators enter (figure 3.14, panel a). This acceleration in the growth of the subscriber base as competition intensifies varies across countries. The impact of competition is felt earlier in the liberalization process in higher-income countries than in lower-income countries (figure 3.14, panel b).

The effect of the introduction of competition on prices is complex. A country's income is a primary determinant of market performance: Higher-income countries have better network coverage and subscriber penetration than lower-income countries, despite higher prices (figure 3.15, panel a).

Sector performance is also affected by policy decisions, particularly the degree of competition in the market. Highly competitive markets

Figure 3.13 Telecommunications Revenue as Percentage of GDP in Selected African Countries

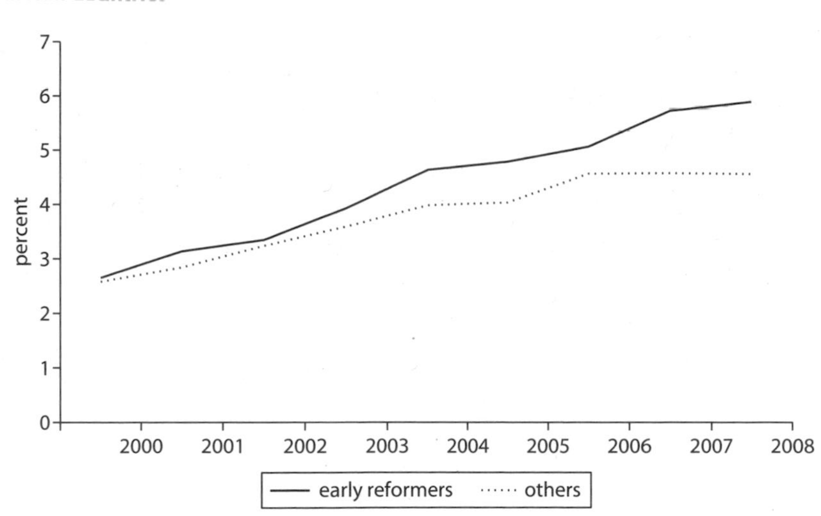

Source: Ampah and others 2009, updated.

Figure 3.14 Impact of Competition on Subscriber Growth

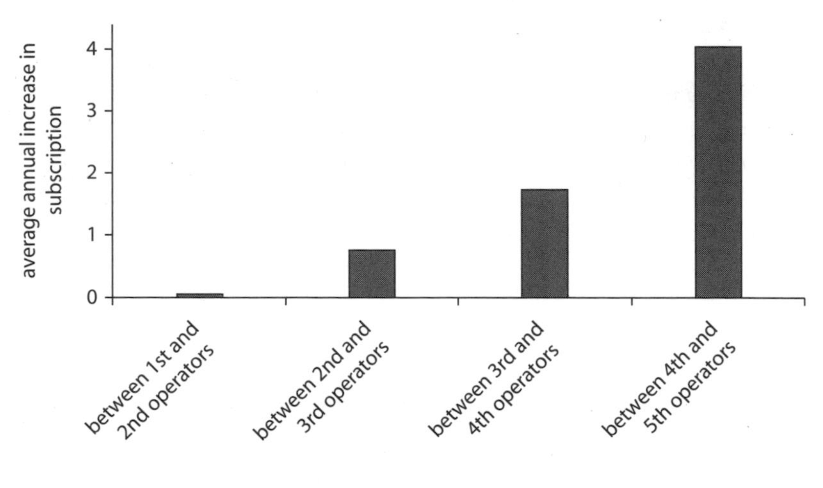

b. Annual increment gains in subscription between entries of telephone operators by level of income (aggregates over the sample)

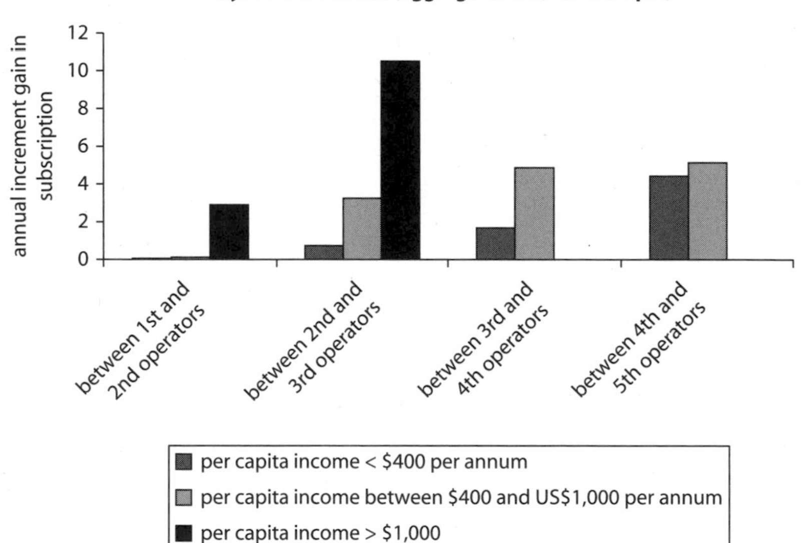

Source: Ampah and others 2009, updated.

(defined by a Herfindahl-Hirschman Index [HHI] of below 5,000) tend to have lower prices, better coverage, and more subscribers than less competitive ones (figure 3.15, panel b).

Overall, prices in the mobile market are falling across the region as competition develops. Following the entry of the second mobile operator, prices fall quickly and converge across countries (figure 3.16).

Despite the overall downward trend in prices as a result of competition, countries with the least-competitive mobile markets tend to have

Figure 3.15 Impact of Country Income and Sector Competition on Mobile Performance, 2008

a. Relationship between country income and mobile sector performance, 2008

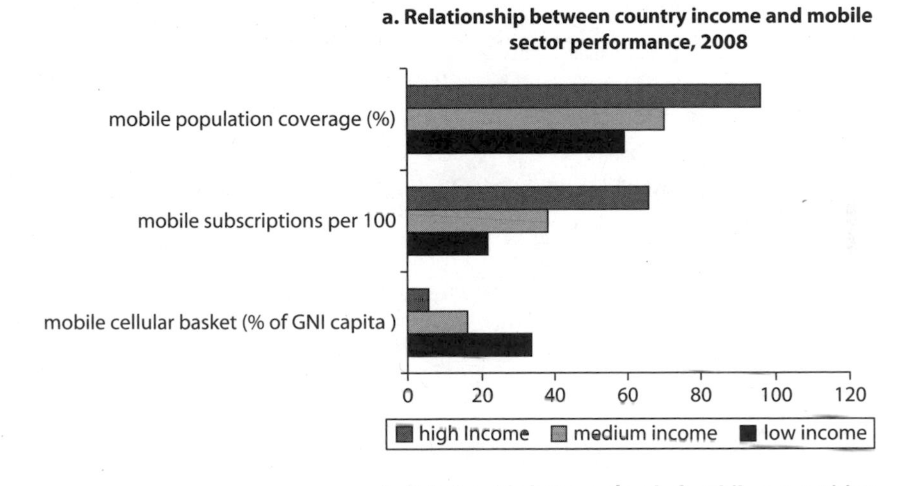

b. Relationship between level of mobile competition and performance, 2008

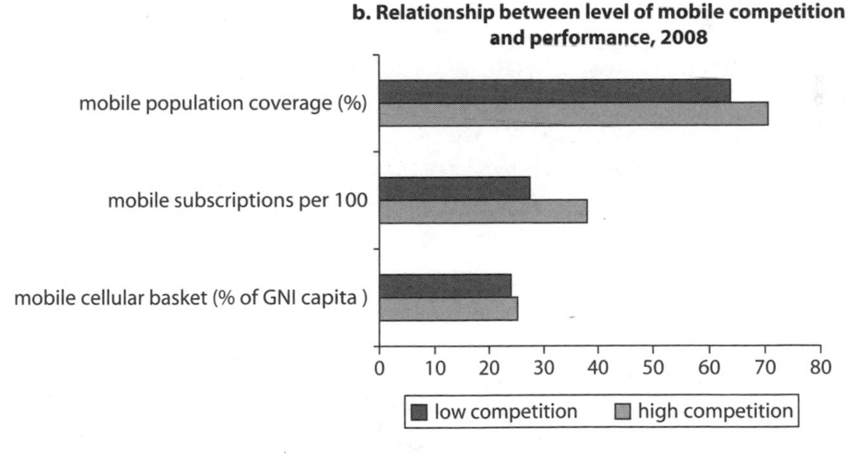

Source: Ampah and others 2009, updated.
Note: GNI = gross national income.

Figure 3.16 Mobile Price Changes after Second Operator Enters the Market

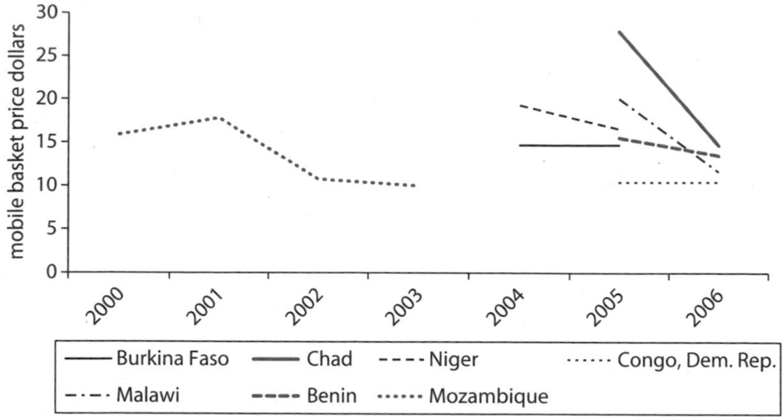

Source: Ampah and others 2009, updated.

Table 3.4 Average Mobile Monthly Price Package and the Herfindahl-Hirschman Index (HHI), 2007

	HHI				
	10,000	7,500–9,999	6,000–7,500	4,500–6,000	3,000–4,500[a]
Mean ($)	12.96	11.17	16.20	15.13	12.08
Mean for low-income countries ($)	10.82	11.17	16.20	15.56	11.97

Source: Ampah and others 2009.
a. Minimum HHI for the sample equals 3,022.

the lowest prices because these operators are owned by the state and are therefore either implicitly or explicitly subsidized. An interesting fact is these countries also typically have lower penetration rates, which indicates that a supply constraint exists (that is, there are people who want to buy services at the prevailing prices but are unable to obtain them). This is usually the result of limited access to finance for network expansion and operational expertise. Prices are highest in the liberalized markets where the market remains concentrated. As competition develops and the market becomes less concentrated, prices fall (table 3.4).

Competition and access to high-bandwidth infrastructure are both prerequisites for cheaper international services, as can be seen in the prices of international telecommunications services across Africa. Countries with competitive international facilities have lower prices

Table 3.5 Price of International Services in Countries with and without Access to Submarine Cables in 24 AICD Countries, 2007

	Percentage of countries	*Price per minute of call within Sub-Saharan Africa ($)*	*Price per minute of call to the United States ($)*	*Price of 20-hour per month dial-up Internet access ($)*	*Price of ADSL broadband Internet access ($)*
No access to submarine cable	67	1.34	0.86	67.95	282.97
Access to submarine cable	33	0.57	0.48	37.04	110.71
Monopoly on international gateway	16	0.70	0.72	37.36	119.88
Competitive international gateways	16	0.48	0.23	36.62	98.49

Source: Ampah and others 2009.
Note: ADSL = asymmetric digital subscriber line; AICD = Africa Infrastructure Country Diagnostic. The table reports data for the 24 countries of phase 1 of the AICD study: Benin, Burkina Faso, Cameroon, Cape Verde, Chad, Côte d'Ivoire, the Democratic Republic of Congo, Ethiopia, Ghana, Kenya, Lesotho, Madagascar, Malawi, Mozambique, Namibia, Niger, Nigeria, Rwanda, Senegal, South Africa, Sudan, Tanzania, Uganda, and Zambia.

than those that retain a monopoly; countries that also have access to submarine cables (such as SAT-3) see even bigger price reductions (table 3.5).

Notes

1. The Comoros has since issued a second mobile license.
2. http://www.ucc.co.ug/licensing/pipPSP.php.
3. This situation changed in mid-2010 with the enactment of the new ICT legislation, the privatization of Zamtel, and the reduction in the price of international gateway licenses.
4. See chapter 2 for a discussion about the definition of fixed and mobile networks.
5. Country-by-country information on the termination of exclusivity for incumbent operators between the key years of 2002 and 2007 is reported in appendix table A2.3.
6. Cape Verde is one example where the government renegotiated exclusivity commitments contained in the incumbent's concession (afrol News and A Semana 2006).

7. See government of Namibia press release: "Namibian Government Lifts Telecom Namibia's Switch Restriction" (http://www.telecom.na/index.php/media/news/2-namibian-government-lifts-telecom-namibias-switch-restriction).
8. Worldwide Interoperability for Microwave Access.
9. In one case, VoIP operators in Namibia were arrested on the grounds of operating an unlicensed telecommunications service. An article about the arrest notes the impact of VoIP on "the viability of Telecom's network by not paying Telecom Namibia right compensation. Illegal Net telephone operators are a pivotal challenge confronting not only Telecom Namibia but also the Namibian Government as a whole. These setups not only put the company's rate/price structure at risk, but drastically reduces the tax benefits that could be reaped by the Government" (ITU 2006).
10. The HHI is "a commonly accepted measure of market concentration. It is calculated by squaring the market share of each firm competing in the market and then summing the resulting numbers. For example, for a market consisting of four firms with shares of 30, 30, 20, and 20 percent, the HHI is 2,600 ($30^2 + 30^2 + 20^2 + 20^2 = 2{,}600$)" (U.S. Department of Justice and Federal Trade Commission 2010, pp. 15–18). Therefore, markets that are more concentrated by being dominated by one or two large players, for example, have a higher HHI than markets in which the players are evenly sized.
11. There are also four full-mobility CDMA operators in Nigeria, further intensifying competition in the mobile market.
12. This situation changed with the government's decision in 2010 to give Zamtel control over ZESCO's fiber telecommunications business in preparation for privatization.
13. The three major satellite operators supplying bandwidth to the African market are Intelsat, Arabsat, and Satellite Communication Services. In addition, there are other smaller operators including Nilesat and Eutelsat.
14. E1 is the international standard unit for small amounts of bandwidth and is approximately equal to 2 megabits per second.
15. An STM-1 is a standard large unit of bandwidth equivalent to 155 megabits per second.
16. Details on privatizations between 1993 and 2009 are reported in appendix 2A.6.
17. Although not strictly a strategic investor, Etisalat, the incumbent operator in the United Arab Emirates, owns 4.6 percent of Sudatel's shares as well as the majority stake in Canar, the second fixed-line operator.
18. For example, in 2003, Pentascope was awarded a three-year contract to manage NITEL, the Nigerian incumbent telecommunications operator. The contract ended prematurely with Pentascope attributing a lack of success due

to "official meddling, sabotage by vested interests that profit from NITEL's inefficiency and fraud" (Nigerian Communications Commission 2005, p. 45). In Tanzania, the government terminated the contract held by Sasktel to manage TTCL, allegedly because of the failure of Sasktel to inject sufficient funding into the incumbent operator (Edwin 2009).

19. "But the board noted that mere acquisition of a dominant market position was not anti-competitive per se if acquired through efficiencies such as better technology, low operational costs, high turnover due to better innovative marketing techniques, superior branding and highly trained technical staff. 'With the proposed acquisition of the two leading mobile telephone operators in Zambia, there are likely significant synergies to accrue that would be used to develop the telecommunications industry,' Mr Lipimile said. He also said that benefits such as increased investment as well as technical and other economies of scale were likely to accrue to the two operators with other envisaged consumer benefits. Among the consumer benefits to accrue include the efficient and real time inter-network short message system at marginal rates and lower tariffs as a result of lower costs of operations and interconnection fees. Also envisaged are more widespread and reliable international roaming possibilities where subscribers would not need to buy a simcard in each country they visited but could still use their Zambian simcard to communicate" (*Times of Zambia*, July 27, 2005).
20. "Effective regulation that supports sustainable investment requires some independence from political influences, especially on a day-to-day or decision-by-decision basis. The regulatory body must be an impartial, transparent, objective and non-partisan enforcer of government-determined policies by means set out in controlling statutes of the regulator, free of transitory political influences. The regulator should also be independent from the industry that supplies ICT services" (Infodev 2010, http://www.ictregulationtoolkit.org/en/Section.3107.html).
21. Some countries have a system known as "bill and keep" in which operators do not pay interconnection charges to one another. Such interconnection arrangements are usually accompanied by a retail pricing structure in which customers are charged to receive calls. The United States is the most notable example of such an arrangement, but they are rare in Africa.
22. See Government of New Zealand (2010) for a recent example of a decision by a regulatory authority to regulate termination charges.
23. "Due to the time required to develop and implement such a methodology, it would not be feasible or desirable to implement a forward looking LRIC approach within the context of the current dispute" (BTA 2003, p. 33).
24. This follows from the definition of call termination as a separate market and the determination that operators are dominant in call termination in their own markets (European Commission 2003).

25. The Village Phone scheme in Bangladesh was spearheaded by Muhammad Yunus, founder of Grameen Bank and winner of the 2006 Nobel Peace Prize. The Grameen Foundation is involved in both the Rwanda and the Uganda operations. See Telenor, "Deep Admiration for Mohammad Yunus." http://presse.telenor.no/PR/200610/1081182_5.html.
26. See Grameen Foundation, http://www.grameenfoundation.org/sub_saharan_africa/uganda/.
27. ISM radio bands were originally reserved for industrial, scientific, and medical use. Because these applications must tolerate interference, many countries have made them unlicensed where they are used for new communications services such as Wi-Fi.
28. A class license authorizes a group of operators to offer a service: "While individual licences are granted to a single service provider at a time, general authorizations provide authority for a whole *class* of service providers to provide service or operate facilities" (Infodev 2010). The license is often automatic and typically granted immediately upon completion of a request form.
29. The performance measures are based on the indicators used by Bressie, Kende, and Williams (2004).

References

AfricaNext. 2008. "3G, WiMAX, ADSL and the Future of African Broadband." AfricaNext Investment Research, Cambridge, MA.

afrol News and A Semana. 2006. "Cape Verde Telecom Market to be Liberalized." September 25. http://www.afrol.com/articles/21535.

Akoh, Ben. 2008. "The Case for 'Open Access' Communications Infrastructure in Africa: The SAT-3/WASC Cable-Senegal Case Study." Association for Progressive Communications. http://www.apc.org/en/system/files/APC_SAT3Senegal_20080516.pdf.

Ampah, Mavis, Daniel Camos, Cecilia Briceño-Garmendia, Michael Minges, Maria Shkaratan, and Mark Williams. 2009. "Information and Communications Technology in Sub-Saharan Africa: A Sector Review." AICD Background Paper 10, Africa Infrastructure Country Diagnostic, World Bank, Washington, DC.

ARTP (Agence de Régulation des Télécommunications et des Postes). 2006. "Rapport sur le marché des télécommunications au Sénégal en 2006." ARTP, Dakar.

Autorité de Régulation Multisectorielle. 2005. "Décision no. 13 du 2 Août 2005, Conseil National de Régulation (CNR) portant liste des opérateurs dominants." Niamey. http://www.arm-niger.org/arm/fichiers/pdf/decision013.pdf.

Balancing Act. 2007. "Four African Countries Try to Turn Back the Clock by Creating Monopoly International Gateways Again." *Balancing Act* (London, online), no. 340.

Bressie, K., M. Kende, and H. Williams. 2004. "Telecommunications Trade Liberalization and the WTO." Paper presented to the 15th ITS Biennial Conference, "Changing Peoples, Societies and Companies: Telecoms in the 21st Century," Berlin, September 5–7. http://www.ppl.nl/bibliographies/wto/files/6620.pdf.

BTA (Botswana Telecommunications Authority). 2003. "BTA Ruling No. 2 of 2003. Ruling on Interconnection Charges Dispute between: Mascom Wireless (PTY) Limited and Orange Botswana (PTY) Limited." BTA, Botswana.

Business Monitor International. 2010. "Kenya: Safaricom to Begin LTE Technical Trials." Business Monitor International, London.

Cohen, T., and R. Southwood. 2004. "An Overview of VoIP Regulation in Africa: Policy Responses and Proposals." *Balancing Act Africa* (London), June.

Communications Commission of Kenya. 2010. "Determination on Interconnections Rates for Fixed and Mobile Telecommunications Networks, Infrastructure Sharing and Co-Location, and Broadband Interconnection Services in Kenya: Interconnection Determination no. 2 of 2010." Communications Commission of Kenya, Nairobi.

Edwin, Wilfred. 2009. "TTCL Deal—Canadian Firm Now Pulls Out." *East African*, July 6.

Esselaar Steve, A. Gillwald, M. Moyo, and K. Naidoo. 2010. "South African Sector Performance Review 2009/2010." Volume 2, Policy Paper 6, Research ICT Africa, Johannesburg.

Esselaar, Steve, A. Gillwald, and E. Sutherland. 2007. "The Regulation of Undersea Cables and Landing Stations." Link Centre, Johannesburg.

European Commission. 2003. "Commission Recommendation of 11 February 2003 on Relevant Product and Service Markets within the Electronic Communications Sector Susceptible to Ex Ante Regulation in Accordance with Directive 2002/21/EC of the European Parliament and of the Council on a Common Regulatory Framework for Electronic Communication Networks and Services." *Official Journal of the European Union* (May 8), Brussels.

———. 2009. "Commission Recommendation of 7 May 2009 on the Regulatory Treatment of Fixed and Mobile Termination Rates in the EU (2009/396/EC)." *Official Journal of the European Union* (20 May 2009) (L 124/67), Brussels.

Global Insight. 2007. "Benin: Beninois Regulator Withdraws Operating Licenses from MTN and Moov." *Global Insight Perspective*, July 17. http://www.globalinsight.com/SDA/SDADetail9935.htm.

Government of New Zealand. 2010. "Reconsideration Report on Whether the Mobile Termination Access Services (Incorporating Mobile-to-Mobile Voice Termination, Fixed-to-Mobile Voice Termination and Short-Message-Service Termination) Should Become Designated or Specified Services." Commerce Commission, Auckland.

Hamilton, Paul. 2010. "Broadband Network Development in Sub-Saharan Africa." Unpublished paper. Hamilton Research, Bath, England.

Hoernig, Steffen. 2006. "On-Net and Off-Net Pricing on Asymmetric Telecommunications Networks." CEPR Discussion Paper No. 5588, Center for Economic and Policy Research, London.

Horvitz, Robert J. 2008. "Beyond Licenced vs. Unlicenced: Spectrum Access Rights Continua." Social Science Research Network. http://ssrn.com/abstract=1259792.

IDA (Infocomm Development Authority). 2004. "IDA Increases Competition in International Telecoms Services." Press Release, September 10. http://www.ida.gov.sg/News%20and%20Events/20050712175459.aspx?getPagetype=20.

Infodev. 2010. "ICT Regulation Toolkit." http://www.ictregulationtoolkit.org/en/Section.668.html.

ITU (International Telecommunications Union). 2006. "Illegal VoIP Operators Arrested at Walvis Bay over the Weekend." *Regulatory NewsLog* (September 17), Geneva.

———. 2008. *African Telecommunication/ICT Indicators 2008: At a Crossroads*. 8th ed. Geneva: ITU.

———. 2009. "Interconnection Agreements and Charges/Reference Interconnection Offer (RIO)." *ICT Eye*. http://www.itu.int/ITU-D/icteye/Reporting/ShowReportFrame.aspx?ReportName=/TREG/InterconnectionAgreementsAndCharges&ReportFormat=HTML4.0&RP_intClassID=1&RP_intLanguageID=1&RP_intYear=2009.

———. 2010. *Measuring the Information Society 2010*. Geneva: ITU.

Kerrets-Makau, M. 2009. "Benchmark Indicators for African National Telecom Regulatory Authority Websites." In *NRA Websites: Benchmarking National Telecom Regulatory Authority Websites*, ed. Amy Mahan, 37–59. Uruguay: Fundación Comunica/LIRNE NET.

Lewis, Charley. 2010. "Achieving Universal Service in South Africa: What Next for Regulation?" Paper presented at the Asia-Pacific Regional Conference of the International Telecommunications Society on Telecommunications Ubiquity and Equity in a Broadband Environment, Wellington, New Zealand, August 26–28.

Milad, Hanna, and B. Ramarao. 2006. "Cost Optimization for Transmission and Backhaul Technologies." *Accenture*. https://accenture.com.mx/NR/rdon

lyres/D310351C-3D90-4F21-80E6-844315DAAD66/43749/acs_cost_opt_pov8.pdf.

Mobile Leader. 2008. "Next-Gen Nigeria." *Mobile Leader* (spring). CDMA Development Group, Costa Mesa, CA.

MTN. 2006. "MTN Ploughs Back Revenue into CSR and Infrastructure Growth in Nigeria." Corporate Press Release. http://www.mtnonline.com/corporate/press.asp?NewsID=125.

Nigerian Communications Commission. 2005. *Trends in Telecommunications Markets in Nigeria 2003–2004*. Abuja: Nigerian Communications Commission.

———. 2006. "Determination of Interconnection Rate." June 21. http://www.ncc.gov.ng/interconnection/Interconnect%20Rate%20Determination%202006.pdf.

Osuagwu, Prince. 2010. "Twist, Turns in Nitel Sale Cancellation as Expert Appeals for Government Attention." *Vanguard*, May 4.

Ovum. 2005. "Recommendations on Further Liberalization of the Telecommunications Industry of Botswana: A Final Report to the Botswana Telecommunications Authority." Ovum, London.

Shannon, Victoria. 2006. "Communications Regulators 'Converge' with the Times." *International Herald Tribune*, December 10.

Singh, Rajendra, and S. Raja. 2010. *Convergence in Information and Communication Technology: Strategic and Regulatory Considerations.* Washington, DC: World Bank.

TCRA (Tanzania Communications Regulatory Authority). 2004. *Determination on Interconnection Rates among the Fixed and Mobile Telecommunications Networks. Interconnection Determination No. 1 of 2004*. Tanzania: TCRA.

TeleGeography. 2010. "TeleGeography's Bandwidth Pricing Report." Washington, DC: PriMetrica.

The East African. 2007. "CCK Caps Interconnection Charges for Warring Mobile Firms." *The East African* (Nairobi), February 25.

TRAI (Telecom Regulatory Authority of India). 2009. *The Telecommunication Interconnection Usage Charges (Tenth Amendment) Regulations, 2009*. New Delhi: TRAI.

UCC (Uganda Communications Commission). 2010. "RCDF Achievements as at 21-01-2010." Kampala. http://www.ucc.co.ug/rcdf/index.php.

UNCTAD (United Nations Conference on Trade and Development). 2007. "Blue Book on Best Practice in Investment Promotion and Facilitation: Zambia." http://www.unctad.org/Templates/webflyer.asp?docid=8183&intItemID=1397&lang=1&mode=downloads.

U.S. Department of Justice and the Federal Trade Commission. 2010. *Horizontal Merger Guidelines.* Washington, DC: Government Printing Office.

Vagliasindi, Maria, and J. Nellis. 2010. "Infrastructure's Institutional Scorecard." In *Africa's Infrastructure: A Time for Transformation*, ed. Vivien Foster and Cecilia Briceño-Garmendia, 105–23. Washington, DC: World Bank.

Wellenius, Björn, and I. Neto. 2007. "Managing the Radio Spectrum: Framework for Reform in Developing Countries." World Bank, Washington, DC.

Williams, Mark D. J. 2010. *Broadband for Africa: Developing Backbone Communications Networks.* Washington, DC: World Bank.

CHAPTER 4

Financing Telecommunications in Africa

Access to finance is often seen as a constraint on economic development in Africa, but the telecommunications sector appears to have overcome this constraint by accessing a wide range of financing sources to fund the rapid expansion of networks.[1]

Operators and governments in Sub-Saharan Africa are investing heavily in the region's information and communication technology (ICT) sector—about $5 billion a year or 1 percent of gross domestic product (GDP). Private sources account for the majority of capital investment in the sector, although a significant amount of money is invested by operators that remain under state ownership. Official development assistance (ODA) from outside the region is marginal overall.

African capital markets, corporate bond markets, and commercial bank loans all have played key roles in financing investment in the telecommunications sector. Although securities exchanges in Sub-Saharan Africa are generally underdeveloped, telecommunications businesses are relatively well represented in them and have successfully used exchanges to raise investment finance.

Despite the wave of privatization and liberalization of the telecommunications market in Africa, the public sector—both domestic and foreign—continues to play a significant role in financing ICT development.

Some African governments retain ownership of one of the operators and sometimes allow that operator exclusive control over one or more segments of the market. More recently, governments have begun investing in fiber-optic backbone networks, either directly or through state-owned operators. However, this strategy imposes significant economic costs on the country: Assets tied up with state-owned operators cannot be used for other, more productive purposes, and when governments have an investment in the market, a tendency is seen to make policy and regulatory decisions to protect that investment, sometimes to the detriment of customers and long-term economic growth.

Private Financing of ICT Investment

The private sector has invested heavily in ICTs since the end of the 1990s, when the expansion of telecommunications networks in Africa began. This investment has fluctuated from year to year, however, and the amount of investment received by each country has varied enormously.

Private Investment Patterns

Between 1998 and 2008, about $50 billion was invested in telecommunications infrastructure projects involving the private sector in Sub-Saharan Africa.[2] As markets have been liberalized and networks have expanded over time, the rates of investment in the sector have increased (figure 4.1).

Since the end of the 1990s, and particularly after 2006, total investment levels have rapidly increased. When these increases are measured against total GDP, however, the pattern is less clear. A major jump took place at the end of the 1990s, as the first wave of market liberalization took place across the region. Since then, the level of investment has fluctuated but with a generally increasing trend, approaching 1 percent of GDP per year by 2008. Investment in telecommunications infrastructure has therefore consistently accounted for a significant, although small, proportion of Sub-Saharan Africa's total GDP throughout the period.

In addition to year-to-year fluctuations in the total amount of investment in the region, investment in the telecommunications sector has varied enormously among countries (figure 4.2). In general, larger countries have received greater levels of investment in telecommunications infrastructure than smaller ones. South Africa is a large and relatively rich country with an advanced telecommunications infrastructure and has therefore received the most investment. Nigeria's income per capita is

Figure 4.1 Private Investment in ICT Infrastructure in Sub-Saharan Africa, 1998–2008

Sources: World Bank Private Participation in Infrastructure (PPI) database; World Bank 2010.
Note: The figure includes only investment in physical assets and excludes the cost of purchasing a license or a state-owned company.

Figure 4.2 Investment in ICT Infrastructure Projects with Private Participation in Sub-Saharan Africa, 1998–2008, Top 10 Countries

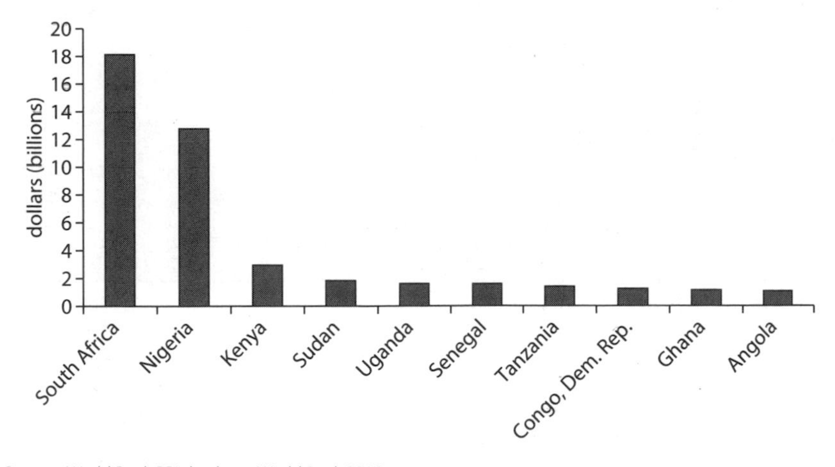

Sources: World Bank PPI database; World Bank 2010.
Note: Figure includes only investment in physical assets and excludes the cost of purchasing a license or a state-owned company. See appendix table A3.1 for full data set.

much lower than South Africa's, but its large population and competitive market structure has resulted in high levels of investment in telecommunications assets by operators.

Although larger countries have tended to receive much greater absolute levels of investment, the amount of investment received relative to GDP has been consistent across countries. For example, figure 4.3 shows the 10 countries in the region with the highest totals of private participation in infrastructure (PPI) in the ICT sector, relative to GDP, over the period 1998–2008 (table 4.1).

Figure 4.3 Investment in ICT Infrastructure Projects with Private Participation in Sub-Saharan Africa as Percentage of GDP, 1998–2008, Top 10 Countries

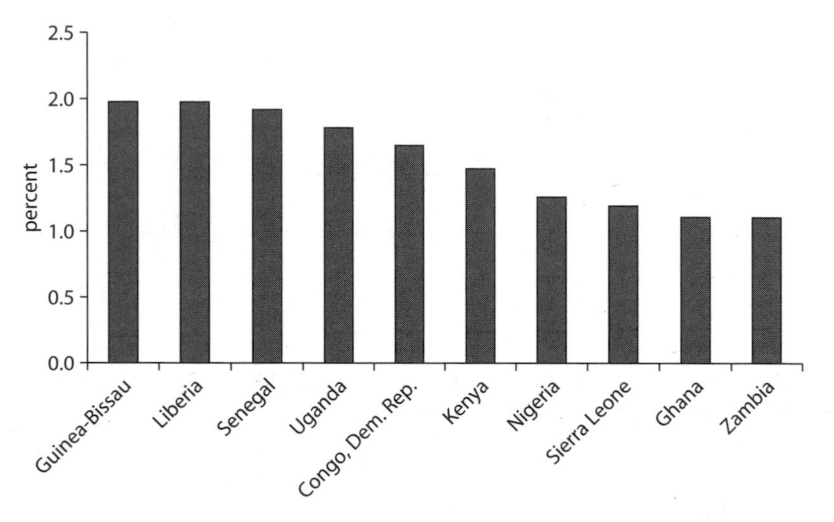

Sources: World Bank PPI database; World Bank 2010.
Note: Figure includes only investment in physical assets and excludes the cost of purchasing a license or a state-owned company. See appendix table A3.1 for full data set.

Table 4.1 Existing Annualized PPI Flows (1998–2008)

Country classification	*Dollars (millions) per year*	*GDP share (percent)*
Sub-Saharan Africa	4,459	0.9
Low-income, fragile	191	0.8
Low-income, nonfragile	935	0.9
Middle-income	1,714	0.8

Sources: World Bank PPI database; World Bank 2010.
Note: Figure includes only investment in physical assets and excludes the cost of purchasing a license or a state-owned company. GDP = gross domestic product; PPI = private participation in infrastructure. The classifications used here are the country classifications used in the Africa Infrastructure Country Diagnostic study (Foster and Briceño-Garmendia 2010).

Four of the countries shown in figure 4.3 are classified as low-income, fragile states (the Democratic Republic of Congo, Guinea-Bissau, Liberia, and Sierra Leone), and four are classified as low-income, nonfragile states (Ghana, Kenya, Senegal, and Uganda); the remaining two (Nigeria and Zambia) are classified as resource rich. Conflict and political instability therefore do not appear to be major deterrents to investments in the telecommunications sector.

The mobile segment of the market has dominated ICT investment in the region. Most of this investment has gone to greenfield projects in which investors either obtain a new license or buy an existing license and then invest in network expansion. Greenfield mobile projects account for approximately three-quarters of the total investment in physical assets in Sub-Saharan Africa, and state-owned enterprises, post-privatization, and greenfield projects in other segments of the market account for the remainder (figure 4.4).

Figure 4.4 Breakdown of Investment in ICT Infrastructure Projects with Private Participation in Sub-Saharan Africa,1998–2008
dollars (billions)

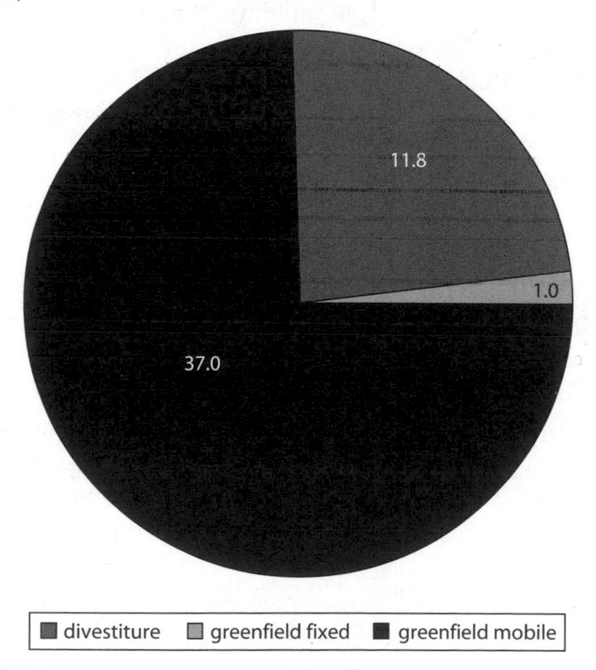

Sources: World Bank PPI database; World Bank 2010.
Note: Figure includes only investment in facilities, not payments made to purchase government assets.

Origins of Private Investment

In the early days of market liberalization, investors from Europe and North America were significant players in the African market. Since then, however, new investors from other regions have played a more prominent role, but there is a distinction to be made between the geographic origin of the sponsor of investments and the source of the finance used to invest. The sponsor of an investment is typically the shareholder with a controlling stake in the business. This sponsor can obtain financing for an investment in a network from a range of sources, including local banks, local securities markets, and international banks. The sponsor's country of origin is therefore often different from the country or countries that provide the investment financing.

Following the initial wave of investment from developed countries in the telecommunications market in Africa, investors from countries outside the Organisation for Economic Co-operation and Development (OECD) have emerged as significant players. For example, companies from the Arab Republic of Egypt, Kuwait, South Africa, and the United Arab Emirates own operators across the region. India has also recently begun to emerge as an important investor in the region through its investments in submarine cables and mobile network operators. Although sponsors of African telecommunications businesses come from around the world, those from developing countries have played a particularly important role in sector investment. Overall, sponsors from Sub-Saharan Africa were responsible for more than half of total investments between 1998 and 2008, sponsors from the Middle East and North Africa accounted for about one-quarter, and those from Europe and North America accounted for less than one-fifth (figure 4.5).

Since 2008, the geographic pattern of investment shows signs of changing. For example, South Asian companies played a very small role in investment in the telecommunications sector in Africa during the period 1998–2008. In 2010, however, it was announced that a major Indian company, Bharti Airtel, would buy the African telecommunications businesses of Zain, one of the biggest and most geographically diverse operators in the region.[3] This purchase may mark the beginning of South Asian investors' ascendancy in the African telecommunications market (*Telecom Finance* 2010).

Another important trend is the emergence of large-scale pan-African operators (see chapter 3 for a more detailed discussion), which have generally been built by companies based in non-OECD countries that have

Figure 4.5 Geographic Breakdown of Sponsors of Telecommunications Businesses in Sub-Saharan Africa, 1998–2008

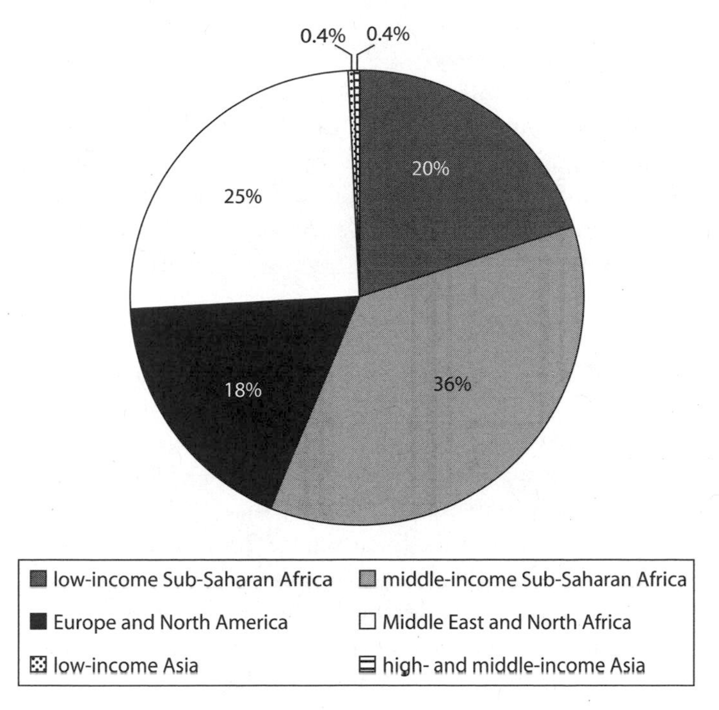

Sources: World Bank PPI database; World Bank 2010.
Note: Data include only cases in which the ownership of the operators is known. Although the major international investors publish data on investment in businesses in other countries, information on local shareholders is harder to obtain. As a result, it is not possible to determine the full ownership structure of many telecommunications businesses. The data therefore likely underestimate the importance of local sponsors in the African telecommunications market.

purchased individual licenses or operators in multiple countries in Africa. In some cases, however, whole groups have been purchased, such as MTC's purchase of the Celtel Group (subsequently rebranded as Zain) in 2005, MTN's purchase of Investcom's businesses in 2006, and Bharti Airtel's purchase of Zain's African businesses in 2010.

Two exceptions are seen to the dominance of non-OECD countries in the emergence of pan-African ICT groups. First, France Telecom invested $1.3 billion in physical telecommunications assets in Sub-Saharan Africa between 1998 and 2008. In addition, Vodafone of the United Kingdom has recently begun to develop a presence in the region, with major investments in Ghana, Kenya, and South Africa. Vodafone also bought Gateway, a major supplier of carrier services in Africa, and used it as a vehicle for expanding its presence in the region.

Commercial Bank Financing

Bank loans are used to finance investment in all types of infrastructure in Africa, and telecommunications infrastructure is no exception. At the end of 2006, outstanding commercial bank loans used to finance infrastructure in Sub-Saharan Africa totaled $11.8 billion. Although it is difficult to determine the exact allocation of these loans among sectors, at least $8.3 billion went to projects in the transport and communication sectors.

Bank loans to the private sector in Africa tend to be short in tenure for all but the most select bank clients, reflecting the predominantly short-term nature of banks' deposits and other liabilities. Financial sector officials in Ghana, Lesotho, Namibia, South Africa, Uganda, and Zambia reported maximum maturity terms of 20 years. Eight other countries reported maximum loan maturities of more than 10 years, and four countries reported 5 or more years. Even where 20-year terms are available, they may not be affordable for infrastructure purposes. In Ghana and Zambia, for example, average lending rates exceed 20 percent, and few infrastructure projects generate sufficient returns to cover such a high cost of debt. The maturities on loans for telecommunications projects vary but are typically about 5 years in length, with an upper limit of 10 years. The average maturity on these loans in Sub-Saharan Africa during the period 2005–09 was 4.9 years, compared with 5.2 years in North Africa, perhaps reflecting a lower risk and a more highly developed commercial loan market in the north (table 4.2).

For most countries in Sub-Saharan Africa, local banking systems are too small and constrained by structural impediments such as poor credit discipline, deficiencies in national legal and judicial frameworks, administrative controls on lending rates, and high transaction costs to adequately finance infrastructural development. Syndicated lending has therefore been an important mechanism for raising financing from local banks for infrastructure projects.[4] The volume of syndicated loans to infrastructure borrowers rose steeply between 2000 and 2006, from $600 million in 2000 to $6.3 billion in 2006. This increase was heavily concentrated in South Africa, which accounted for 80 percent of the total value of syndicated infrastructure loans in Sub-Saharan Africa, and in the telecommunications sector, which accounted for more than three-quarters of the total. Most of the telecommunications sector loans were for projects in

Table 4.2 Syndicated Lending to the Telecommunications Sector

	Total value (millions of dollars)		*Maturity (years) Average (minimum–maximum)*	
	North Africa	*Sub-Saharan Africa*	*North Africa*	*Sub-Saharan Africa*
2005	746.2	220.7	4.3 (0.3–7.0)	3.7 (2.5–5.0)
2006	4,110.8	9,008.7	5.2 (5.0–6.4)	3.8 (1.0–5.0)
2007	3.4	3,901.5	6.0 (6.0–6.0)	6.0 (5.0–7.0)
2008	3,589.4	2,018.0	4.7 (2.1–9.0)	5.1 (1.8–7.0)
2009	29.6	3,347.4	6.0 (5.0–7.0)	6.1 (0.5–10.0)
Total	8,479.4	8,479.4		

Sources: Dealogic; World Bank 2010.

South Africa; the total amount financed outside of South Africa in 2006 was no more than $300 million.

Syndicated loans for telecommunications projects in Africa are typically denominated in local currency to match revenues, and spreads are typically 60 to 100 basis points (table 4.3). Some of these loans, although small by international standards, can be very significant for local markets. Celtel Zambia's loan in 2006, for example, which comprised an $86 million kwacha-denominated tranche raised primarily from Zambian banks and international development finance institutions (DFIs), was, at the time, the largest locally raised kwacha- and foreign-currency-denominated syndicated loan with offshore participation arranged for a Zambian corporation.

Major banks with headquarters in South Africa have played a large part in arranging syndicated deals for borrowers in the telecommunications sector. For example, in 2006, South Africa's Standard Bank provided $27.5 million in financing for the Kenyan telecommunications company Safaricom in addition to the $50.9 million provided by four Kenyan banks. Similarly, in the same year, two South African banks (ABSA Capital and the Development Bank of South Africa [DBSA])[5] and the local affiliate of South Africa's Standard Bank, along with four local banks, a Mauritian bank (Mauritius Commercial Bank), and the international bank Citigroup and the local affiliate of Standard Chartered (UK) participated in the financing arranged for Celtel Zambia.

Two of the biggest debt-financing deals in the telecommunications sector in Sub-Saharan Africa in recent years have been for MTN Ltd. Although MTN is a South African operator, these deals related to

Table 4.3 Details of Syndicated Loans to Telecommunications Borrowers in 2006

	Celtel Zambia	*Safaricom, Kenya*	*Vodacom Group of South Africa*	*MTN Limited of South Africa*
Amount	105	165	1,138	3,468
Currency ($ million)	Local and U.S.	Local	Local	Local and U.S.
Tranches (number)	2	3	1	3
Maturity (years)	5	5	5	3–5
Pricing (basis points)	—	91-day T bill + 100	—	LIBOR + 60–90
Local banks	2	4	4	2
Developing country banks	2 (South Africa)	1 (South Africa)	0	1 (China)
Developed country banks	4	4	1 (Germany)	15

Source: Irving and Manroth 2009.
Note: — = not available; LIBOR = London Interbank Offered Rate.

investments in its businesses in other African countries. In 2006, MTN raised $3.9 billion in commercial loans to finance its $5.5 billion acquisition of Investcom, a Lebanon-based operator with businesses across Africa. A year later, MTN raised a further $1.6 billion in commercial debt financing for investment in its Nigerian business.

Most of the major African telecommunications operators are controlled by companies based in developing countries, but the financing that they have used to invest has tended to come from higher-income countries (figure 4.6). European lenders, in particular, dominated the supply of debt finance over the period 2005–09. This is likely due both to the historical links between Europe's banking sector and Sub-Saharan Africa and to the links between many of the sponsors and European financial institutions. North America, the Middle East, and East Asia have also provided significant financing to the sector. The recent entry of Asian companies into the African telecommunications market may reduce Europe's dominant position, and new sources of debt financing may emerge. By comparison, the financial sector in Africa is relatively underdeveloped, and, as a result, African countries have played a less significant role in providing debt financing, despite their importance as project sponsors (figure 4.6).

Figure 4.6 Geographic Origin of Loans to Telecommunications Operators, 2005–09

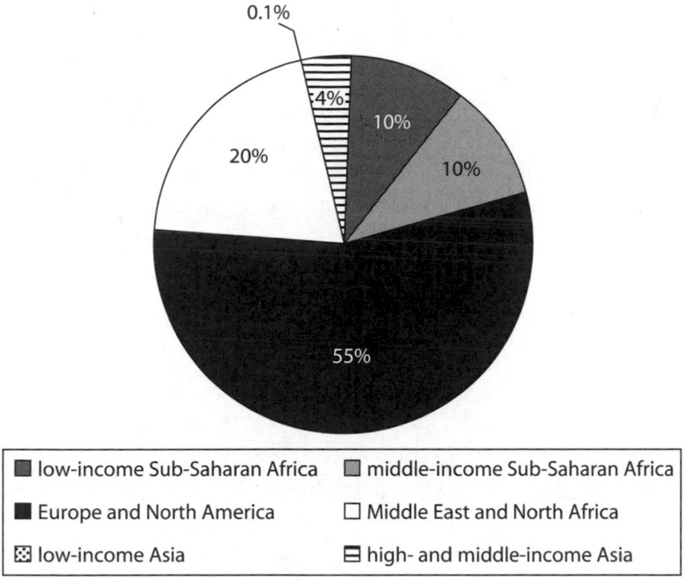

Sources: Dealogic; World Bank 2010.

Financing through Issuance of Securities

Aside from the exchange in South Africa, security exchanges in Sub-Saharan Africa are generally underdeveloped and have played a limited role in the financing of infrastructure investment in the region. Nevertheless, they have supported the development of the telecommunications sector more than other sectors, and in many cases, the financing of telecommunications investments has been an important driver of securities market development in the region.

Equity markets. Most African stock exchanges have very few listings, low liquidity levels, and inadequate market infrastructure. Trading activity is often limited to a few stocks. Total stock market capitalization as a percentage of GDP was less than 33 percent in all the focus countries in 2006 except Kenya (61 percent), Nigeria (33 percent), and South Africa (295.5 percent). It was as low as 0.03 percent in Cameroon, which had a single small listing. In many African countries, most of the local investors who do invest in securities prefer short-term government securities that offer high and liquid returns.

The Johannesburg Securities Exchange (JSE) in South Africa overshadows all other exchanges in the region. At the end of 2006, it had a market capitalization of $621.6 billion, which accounted for nearly 90 percent of the total market capitalization in Sub-Saharan Africa. By comparison, equity listing tallies and market capitalizations for the second- and third-largest stock exchanges—Nigeria and Kenya—were 202 listings and $32.8 billion and 52 listings and $11.4 billion, respectively; the West African Economic and Monetary Union's (WAEMU)[6] eight-member-country regional exchange, the Bourse Régionale des Valeurs Mobilières (BRVM),[7] had a total market capitalization of only $4.2 billion and 40 equity listings, 36 of which were from Côte d'Ivoire.

Infrastructure companies generally do not play a major role in organized national and regional stock exchanges, but there are a few exceptions. Of the 14 stock exchanges that operate in the focus countries, seven (Cameroon, Cape Verde, Ghana, Malawi, Mozambique, Namibia, and Zambia) had no equity listings by infrastructure companies. Furthermore, equity listings by companies operating in infrastructure sectors accounted for only 7.8 percent of the total market capitalization of the JSE (compared with 20.2 percent of South Africa's Bond Exchange), and infrastructure companies accounted for only 0.4 percent of the market capitalization of the Nigerian Stock Exchange.

Although the role of infrastructure companies in the region's stock markets is fairly small, the telecommunications sector is relatively well represented and, in some cases, dominates the local exchanges. For example, equity listings by infrastructure providers accounted for nearly 50 percent of total market capitalization of Sudan's Khartoum Stock Exchange at the end of 2006, which was the highest share in the region. This was nearly entirely composed of a single large $2.3 billion issue by the telecommunications company Sudatel. The BRVM had the next-highest share of infrastructure companies in overall market capitalization, at 47 percent at the end of 2006. Once again, this total was dominated by a single telecommunications company, Sonatel, which accounted for 44 percent of total market capitalization. In South Africa, the telecommunications sector had a total market capitalization of 5.6 percent out of a total market capitalization of 7.8 percent for infrastructure companies as a whole, making it by far the most significant of the infrastructure sectors represented on the JSE.

In addition to listings on local exchanges, two telecommunications firms operating in the infrastructure sector and based in the focus countries were also listed on international exchanges at the end of 2006: South Africa's Telkom SA, Ltd., which was listed on the New York Stock Exchange, and Sudan's telecommunications company, Sudatel, which was listed on the Abu Dhabi and Dubai exchanges in the United Arab Emirates.

Corporate bond markets. In most countries in the region with organized securities markets, the corporate bond market is much smaller than the equity market. For example, according to the Bond Exchange of South Africa, South Africa's corporate bond market was estimated to be only 2–4 percent of the size of its equity market at the end of 2006. Nevertheless, at 13 percent of GDP, South Africa's corporate bond market was by far the largest in the region, with $33.8 billion in issues outstanding at the end of 2006. Outside South Africa, corporate bond markets remain small and illiquid, where they exist at all. The second largest in the region was Namibia's, at $457 million (7.1 percent of GDP). By comparison, corporate bond listing tallies at the end of 2006 for the second- and third-largest stock exchanges—Nigeria and Kenya—were three and eight, respectively, both worth $128 million.

Although corporate bond markets in Africa are very small compared with equity markets, they play a significant role in the financing of investment in the telecommunications sector. Two types of use can be distinguished: (1) for the privatization of state-owned telecommunications incumbents, such as South Africa's Telkom or Sudan's Sudatel, and (2) for the raising of capital for development of new cellular telephone networks.

Overall, $2 billion of corporate bonds issued by telecommunications operators were outstanding at the end of 2006 (table 4.4). As much as $1.9 billion of these were issued in South Africa by Telkom and MTN, which together represented 6 percent of the value of outstanding corporate bonds in that country. Only $100 million in telecommunications bonds were issued outside of South Africa in countries such as Burkina Faso, Kenya, Mozambique, and Uganda. Although the absolute value of these bonds is small, they represent about 12 percent of the value of outstanding corporate bonds in this group of countries—and much more in some cases. A single, exceptionally large listing (Celtel Kenya) accounted for nearly half of total corporate bonds outstanding on Kenya's stock exchange.

Table 4.4 Details of Corporate Bonds Issued by Telecommunications Operators at the End of 2006

	Issuer	*Exchange*	*Issue date*	*Maturity (years)*	*Outstanding value ($ million)*	*Percentage of all corporate bond issues*
South Africa	MTN	BESA	2006	4	717	2
South Africa	Telkom	BESA	1998	10	662	2
South Africa	Telkom	BESA	2005	15	359	1
South Africa	MTN	BESA	2006	8	189	<1
Kenya	Celtel Kenya	Nairobi SE	2005	4	65	51
Burkina Faso	ONATEL	BRVM	2005	6	33	84
Mozambique	MCEL	Maputo SE	2005	5	10	42
Uganda	Uganda Telecom	USE	2003	5	7	22
Burkina Faso	Celtel Burkina Faso	BRVM	2003	6	6	16
Mozambique	TDB	Maputo SE	2004	6	3	13
Total, South Africa					1,927	6
Total, others					124	12
Grand total					2,051	6

Source: Irving and Manroth 2009.

Note: BESA = Bond Exchange of South Africa; BRVM = Bourse Régionale des Valeurs Mobilières; MCEL = Moçambique Celullar; ONATEL = Office National des Télécommunications; SE = stock exchange; TDB = Telecommunication Development Bureau; USE = Uganda Securities Exchange.

Public Financing of ICT Investment

Despite the trend toward privatization and liberalization of the telecommunications market in Africa, governments of about half of the Sub-Saharan African countries still retain ownership of at least one operator. In addition, a limited amount of public investment in the telecommunications sector originates from outside the region, with publicly owned finance institutions providing financing for investment in some telecommunications operators. The government of China plays a significant role in this area through its financing of network equipment supplied by Chinese equipment manufacturers. This is mainly done indirectly through state-owned institutions such as the China Eximbank. Finally, some ODA is also used for investment in the telecommunications sector.

Domestic Public Expenditure in Telecommunications

Governments typically control public expenditure in the telecommunications sector through a state-owned enterprise, where one exists. This expenditure is usually classified as "off-budget," meaning that it is not reflected in the national government budget. Public expenditures in the sector made through the government ministry overseeing telecommunications are "on budget," and the budgets of the regulatory authority may or may not be included in the national government budget. In all, the public sector in Sub-Saharan Africa spends about 1 percent of GDP on telecommunications, including both capital and operational expenditures (table 4.5). This figure is much lower in countries where the incumbent operator has been fully privatized but can reach up to 2 percent of GDP in countries where the incumbent fixed-line operator remains state owned.

Table 4.5 Expenditure Controlled by the Public Sector, 2001–06
annual average, percentage of GDP

Country group	*OPEX*	*CAPEX*	*Total*
Sub-Saharan Africa	0.69	0.26	0.95
Low-income, fragile	0.00	0.00	0.00
Low-income, nonfragile	0.76	0.25	1.01
Middle-income	1.08	0.38	1.46
Resource-rich	0.08	0.08	0.16

Sources: National budgets and operators' annual reports.
Note: Data for the 24 AICD Phase 1 focus countries. CAPEX = capital expenditure; OPEX = operating expenditure.

Off-budget expenditure by state-owned enterprises accounts for about 95 percent of total public expenditure in the sector, but it is important to note that these enterprises also generate significant revenues from charging customers, so the net public expenditure on the telecommunications sector is small. Publicly owned operators in liberalized markets typically have a small share of the market, which likely reflects their constrained access to funds to invest in network expansion and new products.

In contrast to the mobile segment of the market, in which the overwhelming majority of investment originates in the private sector, the backbone segment of the market has offered a more important role for public sector investment (see chapter 3 for a more detailed discussion). The 41,000 kilometers of fiber backbone network that were under construction in Sub-Saharan Africa at the end of 2009 represented a total investment of approximately $800 million. The public sector is funding 49 percent of this amount, either directly or through state-owned enterprises.

Traditionally, all public expenditure on the telecommunications network infrastructure was channeled through a state-owned enterprise, but recent public investment in backbone networks has come from a variety of sources. State-owned operators still play a role, but state-owned electricity companies are also investing, and in some countries the government is investing in fiber-optic infrastructure directly (figure 4.7).

The future of the backbone network infrastructure that has been developed with public funds is unclear at this stage. Some governments have indicated that they intend to turn this infrastructure over to private management and possibly privatize it. Other governments have indicated that they intend to retain ownership of the networks as a national resource.

Public Investment from Outside Sub-Saharan Africa

Governments and other public authorities based outside Sub-Saharan Africa have played a much smaller role in funding the telecommunications sector than they have in other infrastructure sectors. Nevertheless, public investment financing from outside the region has been significant in segments of the telecommunications market and in a few countries.

The telecommunications market is dominated by private companies, and so overseas public financing is usually channeled through the private sector rather than through government-to-government ODA. It also usually originates in public financial institutions such as development banks

Figure 4.7 Investment in Fiber-Optic Backbone Networks in 2009
dollars (millions)

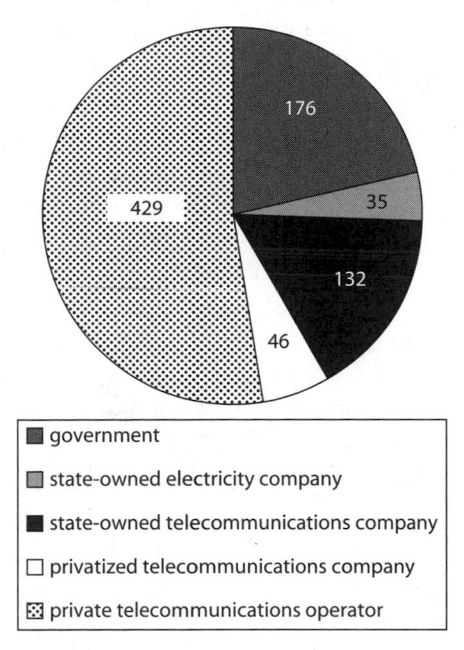

Sources: Hamilton 2010; World Bank 2010.

rather than in ODA budgets, but exceptions to this are found. For example, in 2003, the Organization of the Petroleum Exporting Countries (OPEC) Fund for International Development gave $7.6 million to the incumbent operator in Sudan, Sudatel, to buy digital telecommunications equipment in support of its capital expansion program. LAP Green Networks, an investment company owned by the government of Libya, has invested in the telecommunications sector in Africa and has stakes in numerous operators in the region. In 2007, for example, it acquired a 69 percent stake in Uganda Telecom, the former state-owned operator in Uganda. It then purchased an 80 percent stake in the Rwandan incumbent operator, Rwandatel, following its reprivatization in 2007. Most recently, in 2010 it purchased majority stakes in Gemtel, a mobile operator in southern Sudan, and Zamtel, the incumbent fixed-line operator in Zambia. LAP Green also owns stakes in Sahelcom and Sonitel of Niger and controls Oricel Green, a mobile operator in Côte d'Ivoire.

The government of China is playing an indirect but increasingly important role in the telecommunications sector in Africa. Chinese telecommunications equipment manufacturers supply private operators and governments in Africa with mobile and fiber-optic networks. The three most active Chinese equipment supply firms in the region are the state-owned ZTE Corporation; Huawei, which is privately held; and the French-Chinese joint venture Alcatel Shanghai Bell, in which the private and public sector have an equal 50 percent stake. Many of the projects that these companies undertake are financed through loans from the state-owned China Eximbank or the China Development Bank. Twenty-one countries in Sub-Saharan Africa attracted a cumulative total of almost $3 billion of Chinese public financing for telecommunications between 2001 and 2007, routed mainly through the China Eximbank (figure 4.8). A single project in Ethiopia attracted $1.5 billion for the rollout of a national backbone network and expanded mobile coverage in rural areas. This project was initially agreed upon in 2006, but its implementation has been delayed (appendix table A3.2).

Publicly owned DFIs and other development institutions are a final source of external public financing for the telecommunications sector in Africa. Some of the DFIs, such as the International Finance Corporation (IFC) and the African Development Bank (AfDB), are multilateral institutions. Others, such as Deutsche Investitions-und Entwicklungs-gesellschaft mbH (DEG),[8] the DBSA, and PROPARCO are owned by individual governments. Nonetheless, they typically provide financing on terms similar to that of private investors, so their financing cannot be considered to be concessional in the same way as ODA. It is estimated

Figure 4.8 Chinese Financing Commitments for Confirmed ICT Projects in Sub-Saharan Africa, 2001–07

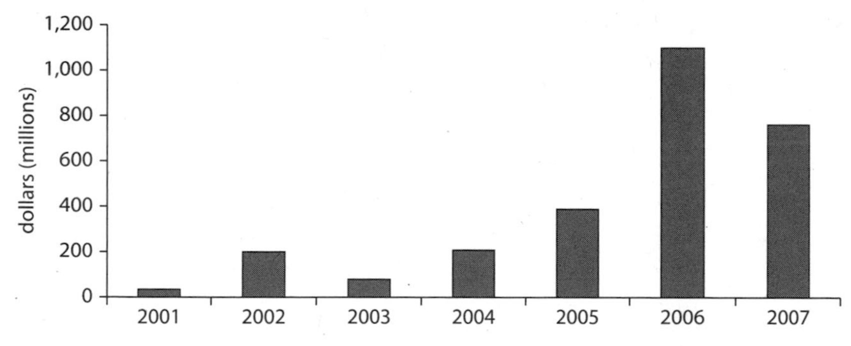

Sources: Foster and others 2008; World Bank 2007.
Note: The data for 2006 include part of $1.5 billion commitment for the Ethiopia Millennium Project.

Figure 4.9 Total DFI Financing of the ICT Sector in Sub-Saharan Africa, 1998–2009

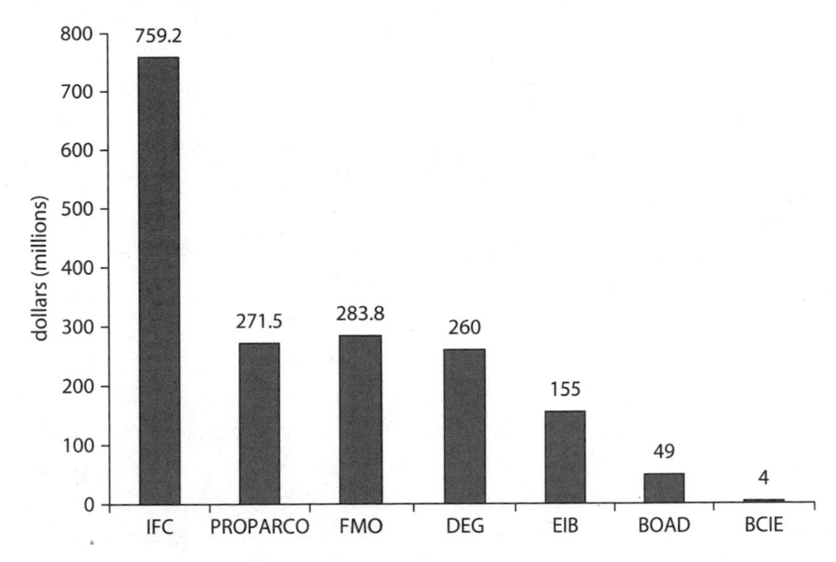

Sources: DFIs; World Bank PPI database; World Bank 2010.
Note: The African Development Bank (AfDB) and the Development Bank of Southern African (DBSA) are believed to be significant sources of finance for the telecommunications sector in Africa, but data on financing provided by these institutions were not available. BCIE = Banco Centroamericano de Integracion Economica; BOAD = Banque Ouest Africaine de Développement; DEG = Deutsche Investitions- und Entwicklungsgesellschaft mbH; EIB = European Investment Bank; FMO = Nederlandse Financierings-Maatschappij voor Ontwikkelingslanden N.V., also known as the Netherlands Development Finance Company; IFC = International Finance Corporation; PROPARCO = Société de Promotion et de Participation pour la Coopération Économique.

that the total amount of financing provided by these institutions to the telecommunications sector in Sub-Saharan Africa between 1998 and 2009 was in the neighborhood of $1.8 billion, although authoritative data were not available for all DFIs at the time of publication (figure 4.9).

Other development institutions are also involved in providing financing for telecommunications infrastructure in Africa. The World Bank, for example, provided $338 million in financing for investment in the ICT sector in Africa between 1998 and 2008. Not all of this, however, was invested in physical infrastructure: It covered a wide range of activities, from policy and regulatory reform to e-government and information technology–industry development. Bilateral ODA for investment in physical telecommunications infrastructure was very limited over the period.

Impact of Public Investment on the Telecommunications Sector

Public sector financing of the telecommunications sector has implications for both its performance and for the public sector as a whole. State-owned telecommunications operators account for the majority of public expenditure in the sector. Because these operators are sustained by user fees and do not require ongoing cash subsidies, they tend to have only a limited direct impact on public budgets. Nonetheless, state ownership of telecommunications operators does impose an economic cost on society, which can be divided into two categories. The first type of cost is the opportunity cost of maintaining a state-owned operator. State ownership ties up expensive public resources in activities that the private sector could take over. By selling these companies, governments could raise significant capital, which could then be used for investment in other areas with higher economic rates of return.

This problem is compounded by the inefficiency of enterprises that are owned by the state. For example, state-owned operators in Sub-Saharan Africa achieved an average of 94 connections per employee compared with the developing-country benchmark of 420 connections per employee; this translated to an overstaffing ratio of 600 percent. State-owned telecommunications companies seem to be particularly plagued by this problem: The overstaffing ratios in publicly owned power and water utilities in Africa over developing-country benchmarks are far lower—88 percent and 24 percent, respectively. Using public resources to pay for this overstaffing—even if it is covered by user fees—diverts funds from more productive uses. If staff numbers were reduced and efficiency improved, the resulting savings could be used for more productive purposes such as investing in network expansion or paying dividends to the government, which could spend the money on another sector. The financial gains from reducing overstaffing in state-owned operators would be significant (table 4.6).

The second type of cost arising from state ownership of telecommunications businesses is due to a smaller, less competitive telecommunications market that, in turn, generates less revenue for the government. Policy

Table 4.6 Potential Gains from Reducing Overstaffing, 2001–06

Country group	*Dollars (millions) per year*	*Percentage of GDP*
Sub-Saharan Africa	1,593	0.33
Low-income, fragile	n.a.	n.a.
Low-income, nonfragile	398	0.36
Middle-income	881	0.33
Resource-rich	314	0.14

Sources: ITU 2008, OECD 2008, operator reports, staff analysis.
Note: n.a. = not applicable.

and regulatory decisions are often skewed in favor of the state-owned operator, which limits competition and indirectly reduces public revenues. The clearest form of this is the maintenance of a monopoly over specific segments of the market, which limits competition and restricts market development. In Zambia, for example, the prohibitively expensive fees set for international gateway licenses maintained the incumbent operator's monopoly until 2010, despite a regionwide trend toward promoting competition in this segment of the market. These high fees resulted in higher prices for international services in Zambia than in other countries, where competition has driven investment and prices have fallen rapidly. Mozambique's state-owned telecommunications operator, Telecomuniçaões de Moçambique (TdM), has also benefited from a de facto monopoly over the provision of national backbone services, despite a general policy of liberalization of the market. As a result, the state-owned electricity operator, which has a functioning fiber-optic backbone network, has not been able to obtain a telecommunications license, which would allow it to compete with TdM. The prices for domestic backbone services charged by TdM therefore remain high, constraining the development of the broadband market.

The stifling effect of state ownership on competition also reduces the tax revenues generated by the sector. Chapter 3 discusses how the telecommunications sector in early reforming countries was bigger, relative to GDP, than in countries that reformed later. Delays in sector reform can therefore translate into reductions in government tax revenues. This can be significant: The telecommunications sector typically generates tax revenues amounting to 25 to 35 percent of the total gross sector revenues (GSMA 2009). Countries that were late to reform their telecommunications sector have therefore been forgoing significant amounts of taxation revenues. This situation is often made worse by regulators and tax authorities treating state-owned operators differently from privately owned operators. In some cases, these operators are not required to pay the same license fees, taxes, and sector levies that private operators pay, either through an official exemption or through failure to enforce legal obligations.

By selling off state-owned telecommunications businesses and increasing competition, governments would increase the overall size of the sector and make it more efficient and competitive, thereby increasing the amount of tax revenues generated by the sector. This could potentially be used to subsidize network rollout into areas that would not otherwise be commercially viable.

Not only does the performance of the telecommunications sector have an impact on tax revenues, but the tax regime also has an impact on the telecommunications sector. In countries in which the telecommunications sector is a major source of public revenue, officials are sometimes tempted to increase tax rates. This can have a negative impact on investment and, ultimately, the growth of the sector. In the Democratic Republic of Congo, for example, the telecommunications sector generates an annual gross revenue of more than $850 million, second only to the mining sector. In 2008, it contributed more than $160 million to the government budget. The telecommunications sector alone represented more than 37 percent of the revenue collected by the national tax collection agency, DGRAD (Direction Générale des Recettes Administratives, Judiciaires, Domaniales et de la Participation), and 30 percent of the tax revenue of the DGE/DGI (Direction des Grandes Entreprises/Direction Générale des Impôts). Seeing the potential for increased tax revenue from the sector, the government adopted a new tax regime specifically for the telecommunications sector in 2008. This came into effect in April 2009 with the imposition of a new excise tax, adding to an already heavy tax burden on the sector.[9] As a result of the new tax, the return on capital in the sector has fallen, and operators have cut back on investment.

Notes

1. Much of the material in this chapter was originally reported in Irving and Manroth (2009). That paper considered the 24 "focus countries" that were included in Phase 1 of the Africa Infrastructure Country Diagnostic: Benin, Burkina Faso, Cameroon, Cape Verde, Chad, the Democratic Republic of Congo, Côte d'Ivoire, Ethiopia, Ghana, Kenya, Lesotho, Madagascar, Malawi, Mozambique, Namibia, Niger, Nigeria, Rwanda, Senegal, South Africa, Sudan, Tanzania, Uganda, and Zambia.
2. This figure includes only projects in which the private sector was involved. It excludes investment by 100 percent state-owned enterprises, on-budget public investments, and ODA.
3. Bharti Airtel bought Zain's telecommunications businesses in the following countries: Burkina Faso, Chad, the Democratic Republic of Congo, Gabon, Ghana, Kenya, Madagascar, Malawi, Niger, Nigeria, the Republic of Congo, Sierra Leone, Tanzania, Uganda, and Zambia.

4. Syndicated lending occurs when a group of banks and other financial institutions form a group or syndicate to provide financing for a project. This increases the total amount of loan financing that a project can obtain.
5. The DBSA is a DFI owned by the government of South Africa.
6. The WAEMU is also known by its French acronym, UEMOA (Union Économique et Monetaire Ouest-Africaine).
7. Its eight member countries are Benin, Burkina Faso, Côte d'Ivoire, Guinea-Bissau, Mali, Niger, Senegal, and Togo. It is headquartered in Côte d'Ivoire but has trading floors in each member country.
8. DEG is part of the KfW Bank Group.
9. Decrees 005/CAB/MIN/PTT/2009 and 006/CAB/MIN/PTT/2009, Journal Officiel de la République Démocratique du Congo, published March 25, 2009.

References

Dealogic. http://www.dealogic.com.

Foster, Vivien and Cecilia Briceño-Garmendia. 2010. *Africa's Infrastructure: A Time for Transformation*. Washington, DC: World Bank.

Foster, Vivien, William Butterfield, Chuan Chen, and Nataliya Pushak. 2008. "Building Bridges: China's Growing Role as Financier for Africa's Infrastructure." Public-Private Infrastructure Advisory Facility, World Bank, Washington, DC.

GSMA (GSM Association). 2009. *Taxation and Growth of Mobile in East Africa*. London: GSMA.

Hamilton, Paul. 2010. "Broadband Network Development in Sub-Saharan Africa." Unpublished paper, Hamilton Research, Bath, England.

Irving, Jacqueline, and Astrid Manroth. 2009. "Local Sources of Financing for Infrastructure in Africa: A Cross-Country Analysis." Policy Research Working Paper 4878, World Bank, Washington, DC.

ITU (International Telecommunications Union). 2008. *African Telecommunication/ICT Indicators 2008: At a Crossroads*. 8th ed. Geneva: ITU.

OECD (Organisation for Economic Co-operation and Development). 2008. *Broadband Growth and Policies in OECD Countries*. Paris: OECD.

Telecom Finance. 2010. "India's African Stampede: Too Late?" *Telecom Finance* (London) no. 177. March.

World Bank. 2007. PPIAF Chinese Projects Database. World Bank, Washington, DC.

———. 2010. *World Development Indicators*. Washington, DC: World Bank.

The results are striking. The proportion of Africa's population living within range of mobile wireless networks has increased by 4 percentage points each year since the late 1990s. As of mid-2009, wireless mobile networks covered 61 percent of the population—a figure that continues to increase every year. In an ideal regulatory environment, it will be commercially viable to build networks to cover an estimated additional 31 percent of the population. Only 8 percent of the continent's population lives in areas that would be unprofitable to serve, which this chapter will refer to as the "coverage gap."

The coverage gap exhibits significant cross-country differences. In the Comoros, Mauritius, and the Seychelles, for example, no coverage gap is found—networks already cover 100 percent of the population. The Central African Republic, the Democratic Republic of Congo, Liberia, and Madagascar, on the other hand, all have coverage gaps greater than 25 percent. In these countries, more than one-quarter of the population lives beyond the commercial reach of GSM networks.

A total expenditure of $15.5 billion would be required between 2007 and 2015 to expand basic GSM network coverage to Africa's entire population. Of this, $6.9 billion is for areas that are potentially commercially viable. The total cost of expanding networks to cover the 8 percent of the population that lies outside these areas amounts to $8.7 billion, or about $1 billion per year.

As the rest of the world adopts the Internet as an integral part of everyday life—and its economic benefits emerge—the lack of widespread access to the Internet in Africa is increasingly seen to be a constraint on growth. Is mass-market broadband Internet commercially viable in Africa and, if so, will broadband networks expand in the way that GSM voice networks did? The analysis in this chapter shows that a limited, public Internet-access model based on wireless infrastructure is commercially viable in large parts of Africa. Such a model could bring basic Internet services within reach of approximately 75 percent of the region's population. Providing home access to the Internet—now common in high-income and some middle-income countries—would require much greater levels of investment. Given the revenue that such a service is expected to generate, much of this network coverage would not be commercially viable in the foreseeable future. Based on current wireless broadband technologies and anticipated revenues, penetration rates of 20 percent in urban areas and 10 percent in rural areas throughout Sub-Saharan Africa would likely require a subsidy of $10.5 billion per year, or 1.3 percent of gross domestic product (GDP).

The Cost of Providing Basic Voice Network Coverage

The analysis of universal coverage costs is based on the analytical framework articulated in a World Bank discussion paper by Navas-Sabater, Dymond, and Juntunen (2002) and further developed in a Regulatel/World Bank study by Stern, Townsend, and Monedero (2006). This framework (summarized in figure 5.1) identifies key supply-and-demand thresholds in a market environment.

Figure 5.1 identifies three areas: the existing coverage, the efficient market gap, and the coverage gap. The innermost rectangle represents *existing coverage*—a country's current population coverage. In a market driven by purely economic forces, suppliers initially serve the areas with the highest revenue potential and the lowest cost per subscriber (that is, the most profitable market segments). They then extend service toward the efficient market coverage frontier, which includes all areas that are commercially viable. Beyond this frontier, suppliers will not provide service without public intervention.

The *efficient market gap* is the difference between existing coverage and the coverage that would be commercially viable if competition were fully effective. This includes all areas in which mobile communications services

Figure 5.1 Coverage Gap Analysis Framework

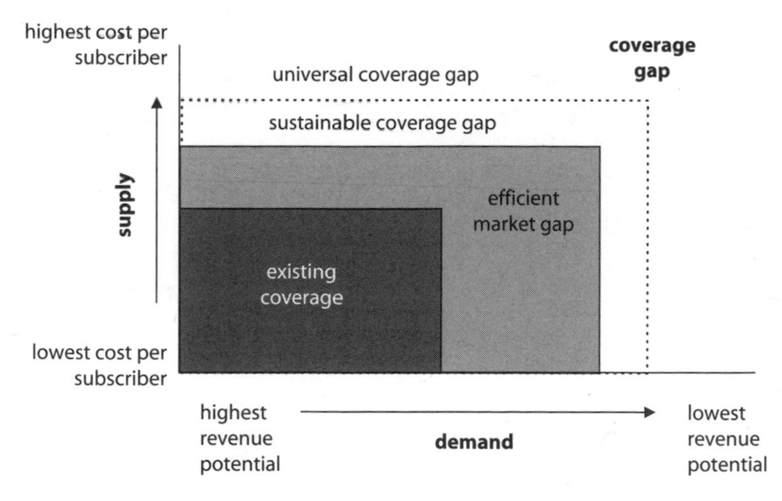

Sources: Mayer and others 2009, adapted from Navas-Sabater, Dymond, and Juntunen 2002, and from Stern, Townsend, and Monedero 2006.

are commercially viable but not currently available. Existing network coverage depends on several factors: the overall costs and the revenues generated by the business, the quality of regulation, the extent of competition in the market, a country's political environment, and the time needed for operators to build out their networks. Potential network coverage depends on the revenue potential and cost of providing coverage in an efficient market in each area of the country. The efficient market gap is measured in terms of either the percentage of the population that could be covered or the cost of infrastructure needed to close the gap.

The *coverage gap* comprises areas where building and operating a network would not be commercially viable. The coverage gap can be divided into two parts: the *sustainable coverage gap* is those areas where revenue potential is high enough to cover operating costs (including a reasonable rate of return) but not to cover initial capital investments over a reasonable depreciation period. The *universal coverage gap* includes all remaining areas that lack sufficient revenue potential to cover either the capital or operating costs of network infrastructure.

Existing Network Coverage

By comparing the geographical distribution of mobile networks in Africa with the distribution of the population, it is possible to estimate the total number of people who are living within range of a network. Since 1999, the proportion of Africa's population living within range of a mobile wireless network has increased by 4 percentage points per year.[2] As of mid-2009, wireless mobile networks covered 61 percent of the population of Africa—up from below 10 percent 10 years before—and this continues to increase. Network coverage in urban areas has consistently been higher than in rural areas. By 2009, 90 percent of Sub-Saharan Africa's urban population was living within range of a mobile network, compared with just under half of its rural population (figure 5.2).

The success of mobile networks in covering the region's population varies enormously across countries. By mid-2009, 10 countries in Sub-Saharan Africa had achieved 90 percent or greater population coverage (Burkina Faso, the Comoros, Gabon, Kenya, Malawi, Mauritius, the Seychelles, South Africa, Swaziland, and Uganda), but 13 countries had less than 50 percent population coverage (Angola, Benin, the Central African Republic, Chad, Eritrea, Ethiopia, Guinea, Guinea-Bissau, Liberia, Mali, Mozambique, Somalia, and Sudan).

Figure 5.2 Proportion of the Population Living within Range of a GSM Mobile Network, 1999–2008

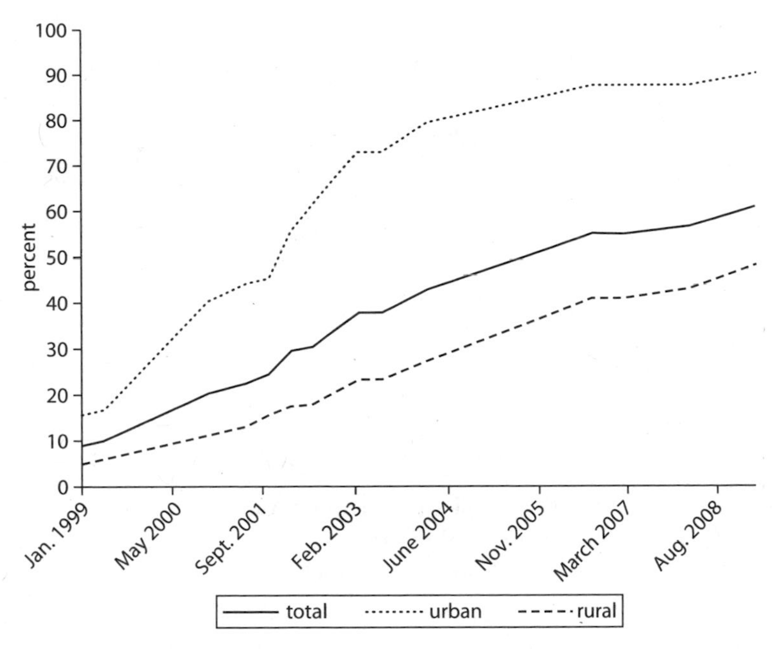

Sources: GSMA 2010b; CIESIN, IFPRI, World Bank and CIAT 2004.

Methodology

The methodology used to estimate the market gaps for the uncovered areas of Sub-Saharan Africa considers four parameters: population density, income distribution, terrain, and size of the wireless cell site. The first two parameters are used to model the revenue potential of voice telephony services in specific geographic regions. The other two parameters—terrain and cell-site size—are used to estimate the cost of providing coverage. In the final stages of the analysis, the estimated costs and revenues for each geographic unit are compared to determine whether a network is commercially viable in that area.

Figure 5.3 illustrates the steps carried out in the spatial analysis for one country, the Democratic Republic of Congo. Populated areas not yet covered by GSM operators are divided into a grid of cells, each the size of a cell site. Each cell is then evaluated in terms of population, terrain, and other characteristics.

For voice infrastructure, GSM was selected as the reference technology. The number of GSM subscribers in Africa far exceeds the totals for

Figure 5.3 Spatial Modeling of Mobile Networks in the Democratic Republic of Congo, 2009

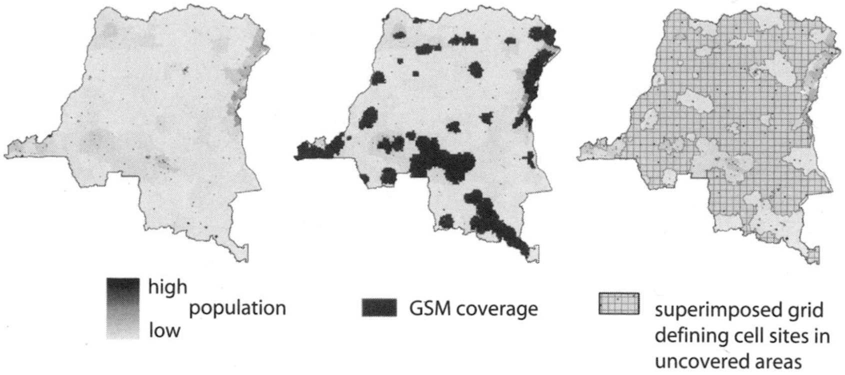

high
population
low
GSM coverage
superimposed grid defining cell sites in uncovered areas

Source: Mayer and others 2009.

any other voice service technology, including traditional wireline systems. Furthermore, GSM is a technology that has achieved significant economies of scale in pricing for both infrastructure and handset products. It is therefore the most cost-effective means of providing wireless signal coverage at this time, with more than 2 billion users worldwide.

Infrastructure investment is divided into capital expenditure (CAPEX) and operating expenditure (OPEX). CAPEX is further divided into the subcategories of network, build, and deploy (NB&D) and network equipment costs (table 5.1). OPEX costs are subdivided into network and fuel costs.

The model was not designed to estimate the cost of providing the infrastructure required to satisfy all market demand. Rather, it determines whether an area has sufficient revenue potential to justify a single base station for providing initial coverage. If it does, but no network currently is present, the area is classified as part of the efficient market gap. If it does not, the area is classified as part of the coverage gap. The model then calculates the investment that would be required to provide the minimum network coverage (that is, a single base station in each cell site) in the area classified as belonging to the coverage gap.

Assumptions and Data

Geo-referenced data were drawn from two primary sources: Estimates of the population distribution within each country were taken from the Global Rural-Urban Mapping Project (GRUMP) data set (CIESIN and others 2004), and historical GSM coverage maps were provided by the

Table 5.1 Voice Services Investment Model: Definitions

Network, build, and deploy (NB&D)

Capital investments that do not include spending on network equipment elements such as radio access, core, and transmission. This category includes cell-site search and acquisition, civil works, and information technology, as well as the design and construction of the site utilities and structures required for the installation of network equipment. Construction management, site design, actual construction work, and all materials including towers, masts, shelters, and power generators are also included.

Equipment capital expenditure (CAPEX)

Capital investments undertaken by the operator in purchasing network-related equipment such as base transceiver systems, mobile switching content, support nodes, servers, switches, routers, and cable (but excluding such nonnetwork equipment items as towers, antennas, and power units).

Radio access network

Spending on the radio access network is defined as investments made to acquire the following units across the TDMA, GSM, CDMA, and UMTS family of technologies: BTS (Node B's), BSCs, and TRXs.

Core access network

Spending on the core access network is defined as investments made to acquire the following units across the TDMA, GSM, CDMA, and UMTS family of technologies: MSCs, HLRs, VLRs, SGSNs, PDSNs, GGSNs, HAs, and AAA servers.

Transmission network

Spending on a carrier's long-haul transport and backhaul network is needed to carry voice and packet communications across large distances. Carriers generally employ the following media for this purpose: copper, wireless (microwave, satellite), and optical fiber.

Network operating expenditures (OPEX)

These include the costs of managing the network operations center, network and service management, performance monitoring, service optimization, and training. They also include the costs of power and maintenance.

Base station fuel costs

Fuel costs represent the cost of diesel fuel to power all cell-site equipment and climate control for one year.

Source: Mayer and others 2009.

Note: AAA = authentication, authorization, and accounting; BSC = base station controller; BTS = base transceiver station; CDMA = Code Division Multiple Access; GGSN = gateway GPRS support node; HA = home agent; HLR = home location register; MSC = mobile switching center; PDSN = packet data serving node; SGSN = serving GPRS support node; TDMA = Time Division Multiple Access; TRX = transceiver; UMTS = Universal Mobile Telecommunications System; VLR = visitor location register.

GSM Association (GSMA 2010b). Key assumptions for model parameters are summarized in table 5.2.

Estimates of revenue potential are based on a percentage of GDP per capita. To calculate an area's total revenue potential, the estimated GDP per capita is multiplied by the revenue potential (that is, the proportion of GDP that is spent on telecommunications services) times the area's population. Total revenue potential is then reduced to take account of value

Table 5.2 Baseline Assumptions Used in the Market-Gap Analysis

Parameter	*Value or definition*	*Source*
Population data	UN forecasts circa 2000	CIESIN and others 2004
GDP	3% growth rate, 2006–15	World Bank task team
Urban share of GDP	GDP share of highest-income percentiles corresponding to percentage of the population that is urban	World Bank 2006
Rural share of GDP	GDP share of lowest-income percentiles corresponding to percentage of the population that is rural	World Bank 2006
GDP per capita (urban)	Urban share of GDP/number of urban inhabitants	Calculation
GDP per capita (rural)	Rural share of GDP/number of rural inhabitants	Calculation
Revenue potential	4% of GDP per capita	World Bank review team
Spatial distribution of population among cell sites	Population distribution and urban/rural characterization supplied by the Global Rural-Urban Mapping Project (GRUMP) alpha data set	CIESIN and others 2004
CAPEX costs	$167,000 per cell site in 2005; declining by an average of 2.1% per year	Mayer and others 2009
Size of cell sites	Rural = 1,662 km^2 Urban = 4 or 8 km^2	Mayer and others 2009
Cell-site capacity	2,000–4,000 subscribers per cell site (typical)	Operator reports
Price of diesel	$0.75 per liter (note: fuel costs are not included in the baseline costing scenario)	Mayer and others 2009
Power consumption of cell sites	3 kW, 24 hours a day, 365 days a year	Mayer and others 2009
Terrain factor	An integer factor ranging from 1 to 4 that is used to adjust the number of base stations per cell site based on terrain. The factor is calculated based on the percentage of raster cells with unobstructed line of sight to a centrally located high point in the cell site representing a hypothetical antenna position.	Mayer and others 2009
GSM coverage	GSM roaming coverage maps	GSM Association 2010b
Cost of capital (PV discount factor)	20%	World Bank review team (with suggested sensitivity analysis of ± 5%)

(continued next page)

Table 5.2 *(continued)*

Parameter	*Value or definition*	*Source*
Network build-out assumptions	The financial analysis models supply-side investment costs based on a scenario in which one-fifth of the remaining network coverage is deployed during each of the first five years of the forecast period (2006–11). OPEX costs and revenues are counted from the first year following the installation of the corresponding cell site through 2015. The costs and revenues are then averaged over the entire period, to produce the present value cost per cell site and revenue potential per inhabitant.	Mayer and others 2009; World Bank review team

Sources: As indicated in table.
Note: CAPEX = capital expenditure; CIESIN = Center for International Earth Science Information Network; km = kilometer; kW = kilowatt; OPEX = operating expenditure; PV = present value; UN = United Nations.

added tax and sector-specific taxes. In the baseline scenario, the assumption for revenue potential was 4 percent of GDP per capita. This was derived from a range of benchmarks. Between 2000 and 2008, telecommunications revenue was equal to about 4 percent of GDP in Sub-Saharan Africa[3] (World Bank 2009). Numerous other surveys cover expenditure on telecommunications services at the household level (table 5.3).

Despite the number of studies on the subject, none reveal a definitive pattern of expenditure on telecommunications in Africa. For example, no general agreement is seen on the relationship between household incomes and expenditure on telecommunications services. Numerous telecommunications demand studies have found that poorer households tend to spend a higher proportion of their incomes on voice services than richer households. Researchers have documented this trend in countries as varied as Chile (Forestier, Grace, and Kenny 2002), India, Mozambique, and Tanzania (Souter and others 2005). These results, however, conflict with the results of broader household income surveys, which show the opposite trend: Telecommunications expenditure as a percentage of household income increases as income rises. Ureta (2005) analyzed spending on telecommunications using household survey data from Albania, Mexico, Nepal, and South Africa. In all four countries, the share of telecommunications expenditures in the average household budget increased with income. Intelecon proposed a third hypothesis, arguing that people spend a certain percentage of their income on telecommunications services regardless of their income (Intelecon 2005). The results of a 2005 survey of telecommunications demand in unserved areas of Nigeria supported this hypothesis.

Table 5.3 Overview of Selected Literature on Household Expenditure on ICT

Source	*Methodology*	*Country/survey year*	*Average budget share spent on ICT*	*Participation rate*[a]
Ureta 2005	Households expenditure surveys Sample population sorted by income deciles and location (urban, rural) Estimated current monthly expenditure on fixed and mobile phone and Internet	Albania/2002–03	3.8%; reaches 4.8% in the tenth decile; urban share exceeds rural share by 0.94% of income[b]	100%
		South Africa/2000	1.4%; reaches 2.5% in the ninth decile; urban share exceeds rural share by 0.99% of income	100%
		Mexico/2000	3.4%; reaches 4.2% in the ninth decile; urban share exceeds rural share by 1.41% of income	100%
		Nepal/2002–03	1.6%; reaches 2.3% in the ninth decile; urban share exceeds rural share by 0.65% of income	100%
Intelecon 2005	Households expenditure survey Unserved areas only Estimated current monthly expenditure on fixed and mobile phones[c] and willingness to pay[d] Missing observations and/or respondents reporting zero expenditure not included in the average	Nigeria/2005	7% currently spent; 0.4% variation between rural and urban households 8% willingness to pay; 0.8% variation between rural and urban households	38% of sample reported current expenditure 91% of sample reported willingness to pay

Souter and others 2005	Households expenditure survey Rural areas only Sample population sorted by income quartiles Estimated current monthly expenditure on fixed and mobile phones	Tanzania/2005	3% on average; reaches 11% in the poorest quartile	100%
		India (Gujarat State)/2005	5% on average; reaches 5.6% in the poorest quartile	100%
		Mozambique/2005	1% richest quartile;[e] 4.2% poorest quartile	n.a.
Moonesinghe and others 2006	Survey localities chosen to capture diversity, not representative sample Survey targeted those earning less than $100 per month who had used a telephone within the past three months Nonusers and higher income levels excluded Estimated current monthly expenditure on fixed, mobile, and public phones Missing observations not included in the average	India (7 localities) and Sri Lanka (4 localities)/2006	Fixed phone: 47% of ≤$50 per month spends less than 2%; 46% spends more than 2% 27% of >$50 per month spends less than 1%; 65% spends more than 1% Mobile phone: 46% of ≤$50 per month spends more than 8%; 45% spends less than 8% 60% of >$50 per month spends more than 4%; 24% spends less than 4%	90%
Gillwald 2005	Desktop study plus household and business survey Estimated average cost of communication (all services included) for users and willingness to pay	10 African countries/2005[f]	10% on average communication costs 5–15% willingness to pay	n.a.

(continued next page)

Table 5.3 *(continued)*

Source	*Methodology*	*Country/survey year*	*Average budget share spent on ICT*	*Participation rate*[a]
Ampah and others 2009	Estimated telecommunications revenue as percentage of GDP	AICD countries/2005	4.07%[g] of GDP	n.a.
Banerjee and others 2007	Household survey Connected areas only (either fixed or mobile or both) Estimated current monthly expenditure Some surveys covered only phones owned by household members. Respondents with both positive and zero expenditure factored in the average Missing observations not specified	Burkina Faso, Chad, the Democratic Republic of Congo, Gabon, Malawi, Mauritania, the Republic of Congo/various years up to 2005	Fixed phone: 19.4% by rural 6.3% by urban Mobile phone: 11.7% by rural 7.03% by urban	2% rural 8% urban 9% rural 35.5% urban

Sources: As indicated in table.
Note: AICD = Africa Infrastructure Country Diagnostic; n.a. = not applicable.
[a] Respondents as percentage of the sample.
[b] Variation resulting from regressing expenditure on location, controlling for sex, education, and age.
[c] For which respondents travel to gain access. Travel costs isolated and excluded from calculated ICT costs.
[d] Maximum share people would pay if service were available.
[e] No average available; expenditure by quartiles only.
[f] Botswana, Cameroon, Ethiopia, Ghana, Namibia, Rwanda, South Africa, Tanzania, Uganda, and Zambia.
[g] Total telecommunications operator revenues/GDP.

Household income and expenditure surveys that are designed and implemented at the national level are typically more representative of telecommunications spending patterns than telecommunications-demand studies. The telecommunications-related questions in standard household surveys, however, are not always specific enough to support accurate market assessments. For example, the Malawi integrated household survey of 2004–05 asked only about spending on telephones owned by the household. In addition, interviewers were directed to skip certain questions based on respondents' answers. A respondent who answered *no* to the question of whether anyone in the household owned a fixed or mobile phone would not be asked any questions about household expenditure on telephony. Therefore, the survey results systematically excluded expenditure on pay phones and phone shops, which are often the main form of telephone access for very low-income populations. In addition, household surveys often failed to distinguish between the spending patterns of respondents in served and unserved areas, thereby producing results that are of limited use in supporting future predictions. The rapid recent increase in the availability of mobile telecommunications services also means that expenditure on these services is often not fully captured in the household surveys that are currently available (Milne 2006).

Given the lack of consensus on the current level of household expenditure on telecommunications services, it is difficult to estimate the revenue potential of areas that lie outside network coverage. The study adopted a relatively simple approach by assuming expenditure on telecommunications services was 4 percent of GDP per capita in all areas of the country, but the estimate of GDP per capita was adjusted to differentiate between rural and urban areas. Using the income distribution curve for each country, the proportion of GDP corresponding to the rural share of the population—starting at the bottom of the income curve—was determined. Urban areas were then assigned the proportion of GDP corresponding to their share of the population, but beginning at the top of the income curve. Average weighted GDP per capita for rural and urban areas was then calculated by dividing rural and urban GDP by the number of rural and urban inhabitants, respectively.

Results

The results of the analysis are striking. Approximately 61 percent of the population of Sub-Saharan Africa live within range of mobile networks, leaving 39 percent of the population living outside of the area of coverage.

Of this unserved population, 240 million people—or 31 percent of the total population—will be provided with network coverage if fully effective competition is established. The simplest and most cost-effective way to increase infrastructure provision for basic voice services in Sub-Saharan Africa is therefore to promote competition in the market.

Only 8 percent of the region's population lives outside the area that is potentially economically viable, or in the "coverage gap." This gap is likely to require some kind of public support if universal coverage goals are to be reached. Results by country are shown in figure 5.4. More detailed results can be found in appendix table A4.1.

The size of these gaps varies widely across countries. In Mauritius, for example, where no coverage gap was found, full competition would likely lead to 100 percent network coverage. In the Central African Republic, on the other hand, 62 percent of the population would remain outside the network coverage, even with fully effective competition. About one-quarter of the countries studied (12 out of 45) have coverage gaps ranging between 10 and 25 percent of the population. Four other countries have a coverage gap of more than 25 percent of their populations: the Central African Republic, the Democratic Republic of Congo, Liberia, and Madagascar. The common features of these countries include low income levels, recent political upheaval, and physically challenging or sparsely populated terrain.

In total, it is estimated that $15.5 billion is needed to provide a minimum level of coverage to those who live out of range of current mobile networks. This figure includes capital investment and infrastructure operation and maintenance. Over a period of nine years, this amounts to an average annual investment of about $1.7 billion. If effective competition is established in all mobile markets, it is expected that the private sector will provide $0.76 billion per year of this. A further $0.96 billion would be needed each year to provide a minimum level of coverage outside commercially viable areas. Detailed results by country are provided in appendix table A4.2.

Although the efficient market gap is four times larger than the coverage gap in terms of population, closing the coverage gap is more costly. The population in the coverage gap is typically the most remote and most unevenly distributed and therefore requires more base stations. As a result, closing the coverage gap involves higher infrastructure cost per person—and the public funding gap rises rapidly as networks extend into more sparsely populated parts of the country. This is consistent with experience in other parts of the world. In Chile, for example, the average

Figure 5.4 Results of Coverage Gap Analysis

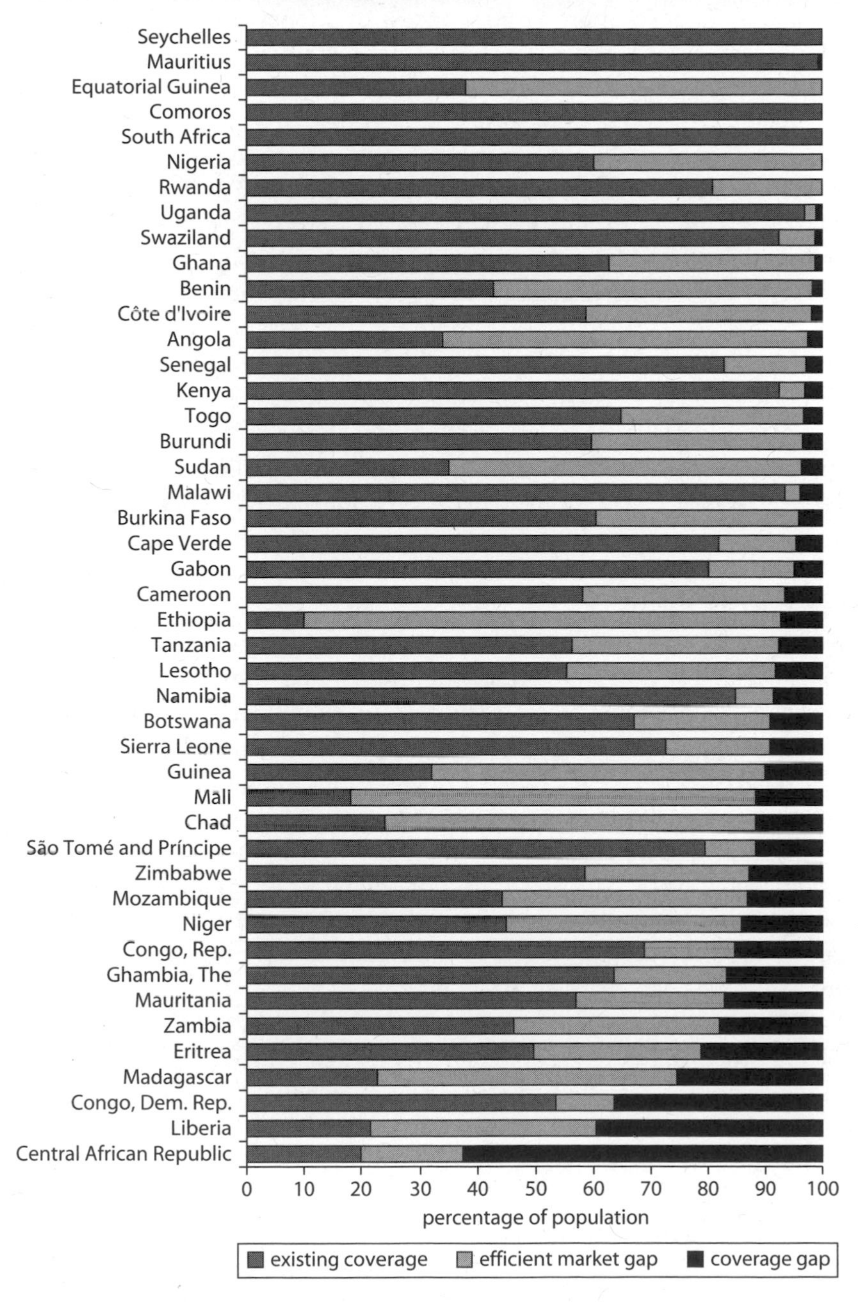

Source: Mayer and others 2009.

Note: The analysis was carried out in the 45 Sub-Saharan African countries for which data were available.

subsidy to provide village pay phones (and to reduce the percentage of the country's population without access to phones from 1.3 percent to 1.0 percent) was 10 times the subsidy per pay phone needed at the start of the program, when the coverage gap was reduced from 15 percent to 9 percent of the population (Wellenius 2002).

The investment needed to provide universal network coverage is substantial, particularly in the low-income countries of Sub-Saharan Africa, but it is important to put this figure into a sectorwide context. It is estimated that the telecommunications industry in Sub-Saharan Africa invests $5 billion per year (see chapter 4) and that the annual total revenue generated by the industry is approximately $39 billion.[4] The approximately $1 billion per year that would be required to provide universal network coverage in the region would therefore cost only about 2.5 percent of the annual revenue generated by the industry. It is also worth noting that the sector generates, on average, about $5 billion in tax revenues per year (GSMA 2010a), which is approximately five times the amount that would be required to fund universal network coverage in the region.

Sensitivity Analysis

These modeling estimates were tested to determine the robustness of the conclusions. The two key input assumptions that were considered were the values for potential revenue and the cost of network construction and operation. The baseline assumption for revenue potential in the model is 4 percent of GDP per capita. If the revenue potential is reduced to 3 percent of GDP per capita, the coverage gap increases from 8 percent to 11 percent of the population (figure 5.5, panel a), and the total public funding gap (2007–15) rises from $6.8 billion to $8.0 billion (figure 5.5, panel b). When revenue potential increases to 6 percent of GDP per capita, the coverage gap drops to 6 percent of the population, and the public funding gap drops to $0.59 billion per year.

Changes in the assumed revenue potential have a more pronounced effect at the country level. For example, Eritrea, Guinea, and Madagascar all show greater sensitivity to changes in revenue potential than most other countries. Figure 5.6 shows the range of the coverage gap as the assumed revenue potential varies from 2 to 6 percent of GDP per capita. The notch in the middle of each range corresponds to the coverage gap for the baseline assumption of 4 percent. For most countries, the length of the range is greater to the right of the notch than to the left. This indicates that the size of the coverage gap is more sensitive to a decrease in revenue potential than to an increase.

Figure 5.5 Sensitivity of Coverage Gap and Public Funding Gap to Revenue Potential for Voice Services (Sub-Saharan Africa)

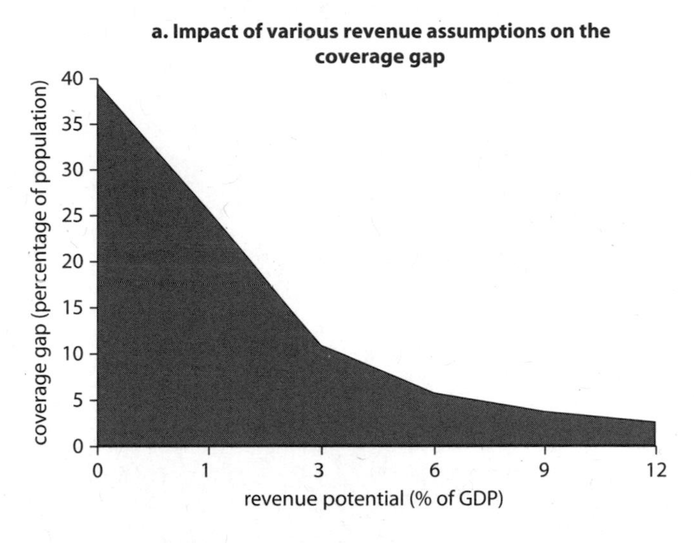

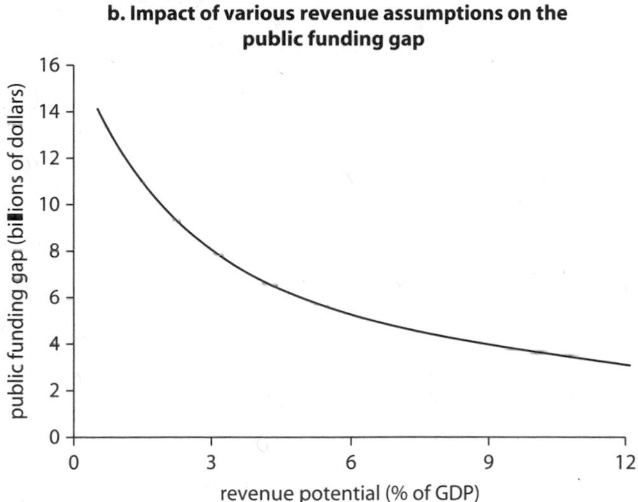

Source: Mayer and others 2009.

The spatial analysis developed for the study relies on assumptions for the area covered by a single GSM cell site. In rural areas, the standard cell-site area was set at a maximum of 1,662 square kilometers (km^2). This is equivalent to a circular area with a radius of approximately 23 km. This radius is well within the functionality of commercially available extended-range GSM base stations, which report coverage radii of up to

Figure 5.6 Country-Level Sensitivity of Coverage Gap to Assumed Revenue Potential

percentage of population in coverage gap as revenue potential varies from 2% to 6% of GDP per capita

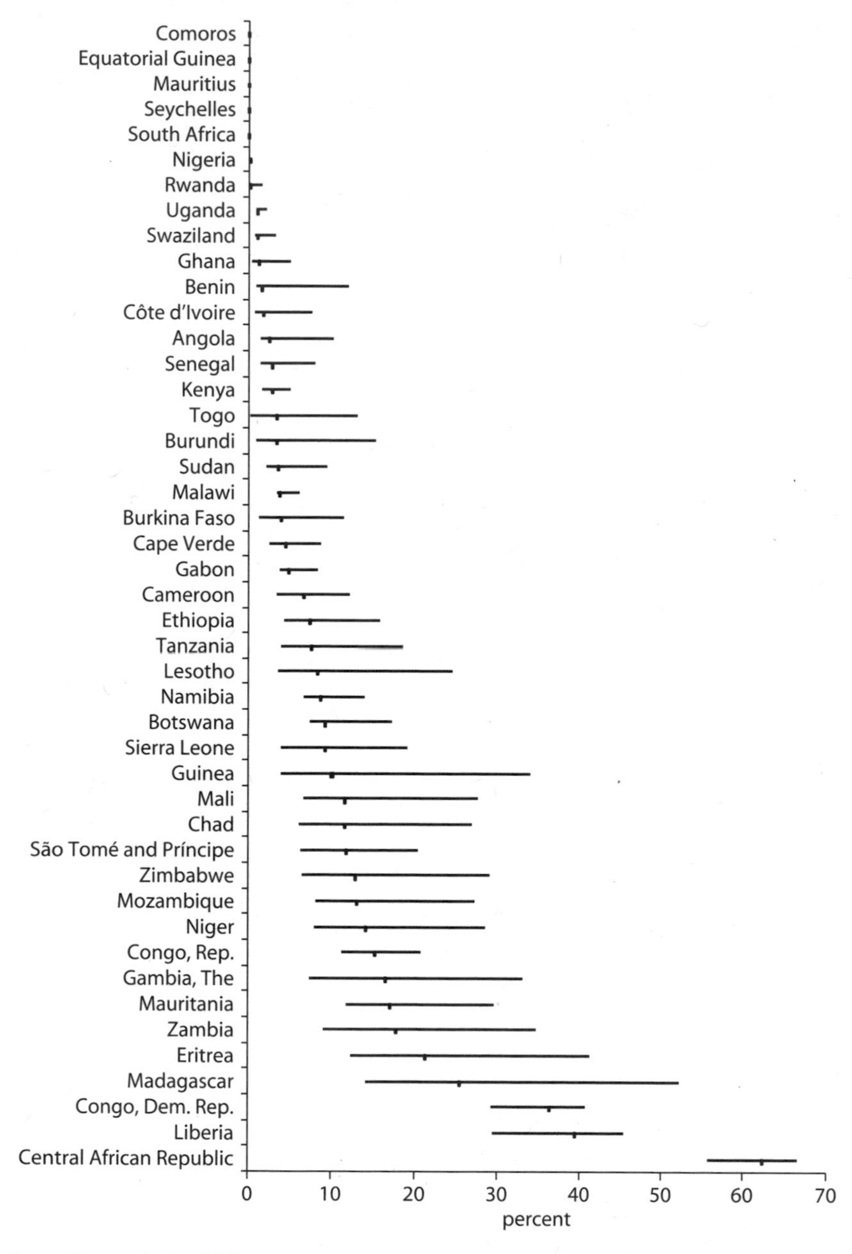

Source: Mayer and others 2009.

50–55 km in flat terrain. In uneven terrain, however, the functional range of GSM base stations is reduced significantly. This was accounted for in the modeling by changing the number of base stations per nominal cell site according to the percentage of the area within the line of sight of a centrally located antenna. The effects of vegetation on signal loss could not be included in the analysis due to lack of data; the digital terrain data used to assess line of sight were coarse, at only 90 meters' resolution. These limitations in the methodology are more significant when the analysis is done at the country level than at the aggregate regional level.

The number of mobile cell sites calculated by a high-level modeling exercise such as this is likely to differ from the number that is needed in practice. Many reasons can be identified. For example, the optimal base station site(s) may not be available for legal or commercial reasons. Small-scale obstructions, such as new building complexes and vegetation, are not captured in this type of aggregate analysis. Meanwhile, operators may have logistical preferences for sites closer to roads and other existing infrastructure rather than at points that provide maximum coverage. For gauging of the sensitivity of the analysis to differences between the theoretical and the actual number of cell sites required to provide coverage, the model was tested with an increased number of sites. Even using an extreme scenario in which the capital cost of cell-site infrastructure was three times that of the base case, the analysis indicates that the proportion of the population living outside commercially viable areas would not rise beyond 20 percent (figure 5.7).

This leads to another striking conclusion. Even if the cost of mobile network investment were significantly higher than it is now, 80 percent of Africa's population would be living in areas that are commercially viable to provide coverage. This means that, even taking a conservative approach, it is reasonable to expect mobile network coverage to continue expanding from the current level of 61 percent population coverage to 80 percent. In reality, however, the cost of mobile network infrastructure is continually falling as GSM equipment prices fall and as operators implement innovative arrangements such as tower outsourcing to reduce costs. It is therefore more likely that, in reality, the networks have the potential to reach the 92 percent population coverage of the baseline scenario.

Despite this positive outcome at the aggregate level, the analysis shows significant variations across countries. Guinea and Madagascar, for example, show high sensitivity to increased infrastructure costs. In Chad, a threefold increase in infrastructure costs shifts 29 percent of the

Figure 5.7 Country-Level Sensitivity of Coverage Gap to Infrastructure Cost Assumptions

percentage of population in coverage gap when infrastructure costs range from one to three times the study's assumption

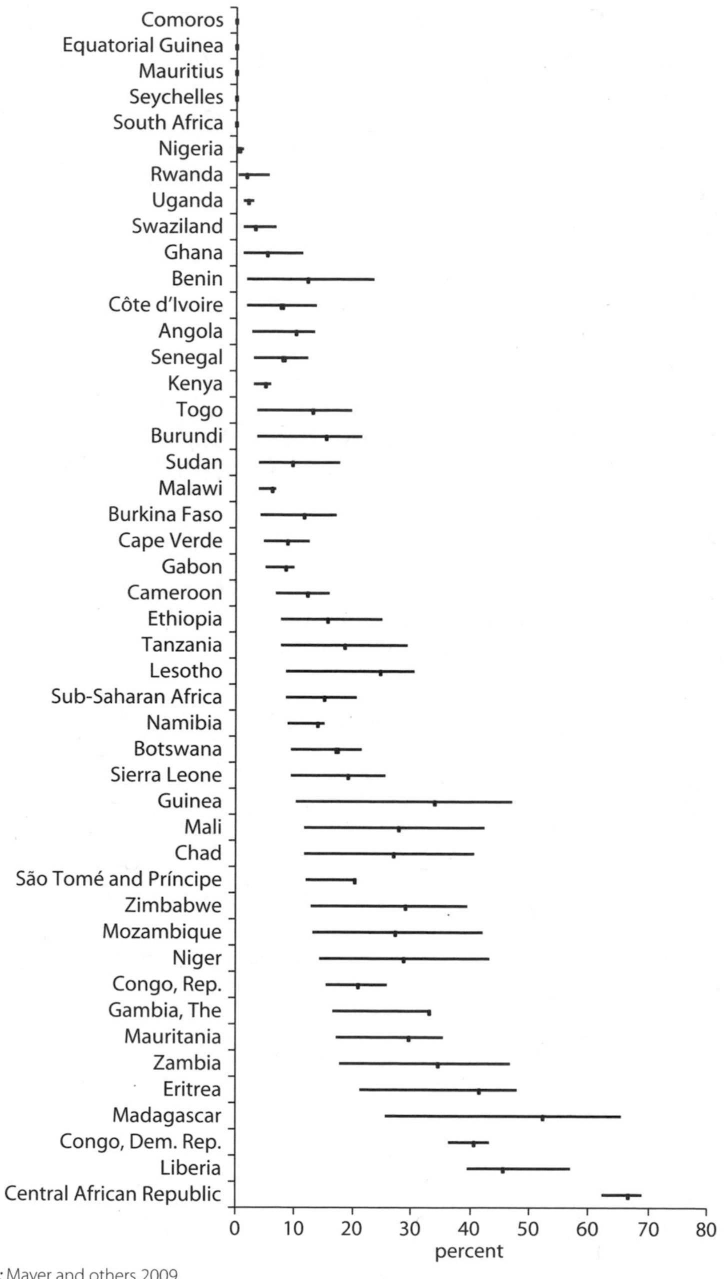

Source: Mayer and others 2009.

population into the coverage gap. Eritrea, Mali, Mozambique, Niger, Zambia, and Zimbabwe experience similar shifts when infrastructure costs increase. In other countries, a tripling of infrastructure costs has a fairly modest impact on the coverage gap. This is to be expected in countries that are close to achieving universal coverage, such as Kenya, South Africa, and Uganda. Nigeria appears to be a unique case in being largely immune to higher infrastructure costs. Even the most extreme rise in infrastructure costs—six times higher than the baseline scenario—increases the coverage gap in Nigeria to only 1.4 percent of the population.

Meanwhile, the cost of providing universal coverage beyond commercially viable areas (that is, the cost of filling the coverage gap) is much more sensitive to assumptions about infrastructure costs. When CAPEX and OPEX triple, the investment needed to close the coverage gap increases by 4.4 times, whereas the investment needed to close the efficient market gap increases by only 1.2 times. Public funding requirements increase by a factor of five—from $6.8 billion to $34 billion (table 5.4). This is due to the shift of some cell sites from the category of commercially viable (efficient market gap) to nonviable (coverage gap) as costs increase but revenues are held constant. In those countries

Table 5.4 Sensitivity of Public Funding Gap to Cost Assumptions
dollars (millions)

	Best estimate of public funding gap	*2 × cost assumptions*	*3 × cost assumptions*
Angola	143	741	1,571
Benin	19	111	331
Botswana	110	392	672
Burkina Faso	47	267	496
Burundi	7	39	85
Cameroon	187	613	1,101
Cape Verde	12	31	54
Chad	119	490	996
Comoros	0	0	0
Congo, Dem. Rep.	1,807	3,929	5,991
Congo, Rep.	272	656	1,031
Côte d'Ivoire	33	160	381
Equatorial Guinea	0	0	0
Eritrea	75	208	356
Ethiopia	372	1,226	2,189
Gabon	84	224	375
Gambia, The	12	54	90

(continued next page)

Table 5.4 *(continued)*

	Best estimate of public funding gap	*2 × cost assumptions*	*3 × cost assumptions*
Ghana	28	128	330
Guinea	53	249	503
Kenya	161	470	767
Lesotho	9	62	106
Liberia	86	202	320
Madagascar	364	1,120	1,994
Malawi	49	106	165
Mali	367	986	1,818
Mauritania	153	391	651
Mauritius	0	0	0
Mozambique	329	965	1,681
Namibia	36	105	178
Niger	166	480	820
Nigeria	38	125	245
Rwanda	5	12	34
São Tomé and Príncipe	4	10	18
Senegal	44	158	311
Seychelles	0	0	0
Sierra Leone	21	92	185
South Africa	29	61	92
Sudan	403	1,105	2,081
Swaziland	10	27	58
Tanzania	259	793	1,441
Togo	5	46	109
Uganda	56	130	216
Zambia	323	1,164	2,056
Zimbabwe	102	365	671
Sub-Saharan Africa	6,794	19,340	33,881

Source: Mayer and others 2009.

that are more sensitive to infrastructure costs, a combination of population density, income, and terrain characteristics result in a greater percentage of the population living in areas that are very close to the threshold of commercial viability.

For a check of the validity of the model's estimates, the model predictions were compared with the actual coverage of GSM operators in rural areas in 23 of the study countries. A summary of the results is shown in table 5.5. In most cases, the model predicted 70–100 percent of actual coverage.

Table 5.5 clearly shows that the model tends to underpredict the extent of the commercially viable network coverage. Various reasons can

Table 5.5 Percentage of Existing Rural Coverage Predicted by the Model

	Rural Coverage
Benin	73
Burkina Faso	88
Cameroon	70
Chad	79
Congo, Dem. Rep.	46
Côte d'Ivoire	74
Ethiopia	58
Ghana	86
Kenya	83
Lesotho	47
Madagascar	60
Malawi	71
Mozambique	61
Namibia	34
Niger	60
Nigeria	90
Rwanda	100
Senegal	74
South Africa	91
Sudan	86
Tanzania	68
Uganda	84
Zambia	73

Source: Mayer and others 2009.

be given. First, because the model uses population maps to determine revenue potential, it could not correctly predict such potential in areas that might have high mobile traffic but little residential population, such as major highways and business and industrial areas. The baseline model parameters were also deliberately chosen to be conservative estimates of revenue potential, infrastructure costs, and the benefits (for the operator) of having broad geographic coverage. In addition, the method for dividing geographic areas into cells to estimate coverage may have introduced errors. Actual network deployments take into account terrain and surface features at a much finer level of resolution than is possible to do when looking at the regionwide level. Real deployments also optimize cell coverage to incorporate the population more efficiently, thereby increasing the commercial viability of some areas. The conclusions of this modeling exercise should therefore be considered as a conservative estimate of the likely extent of network coverage in the region.

Wireless Broadband Infrastructure

Estimating broadband network infrastructure rollout follows a similar approach to that of the voice network model. The investment and operational costs of providing wireless broadband network coverage are estimated using a spatial analysis of the drivers of network costs, based on spatial population and geo-type data. Costs are then compared with potential revenues from broadband services, providing an estimate of the areas in each country in which wireless broadband is commercially viable.

Within this broadly similar framework, significant differences are nevertheless found between the broadband and the voice infrastructure models in terms of both cost and revenue formulas. The major differences on the cost side are the technologies deployed and the inclusion of international bandwidth costs in operating expenses. Internet services generate more international traffic than voice services; the price of international connectivity is therefore a more important cost driver for broadband. The major difference on the revenue side is the development and comparison of two retail service scenarios: one dominated by shared access to broadband facilities, and the other dominated by individual subscriptions. The demand scenarios differ in terms of the number of broadband lines provisioned relative to the population served and how revenues are modeled.

The purpose of developing two demand-side scenarios was to compare the extent of the efficient market at the extremes of prevailing broadband access models: shared and individual access. These poles should provide reference points to policy makers, who will need to balance policies to promote competitive broadband infrastructure provision with universal access objectives.

Spatial Approach to Estimating Wireless Broadband Infrastructure Needs

To estimate the number of base stations needed, two technologies were considered: CDMA2000 1× EV-DO in the 450 megahertz (MHz) band for rural areas (Ho 2005) and WiMAX (rev 802.16-2004) in the 3.5 gigahertz (GHz) band for urban areas. The technologies do not have identical speed or mobility, but they both meet standard definitions of broadband (see chapter 2) and are commonly deployed types of broadband network in Sub-Saharan Africa. Box 5.1 provides more information regarding the assumptions underlying the spatial analysis; appendix

Box 5.1

Wireless Broadband Infrastructure Basics

Third generation (3G) wireless broadband networks are based on a cell architecture similar to mobile wireless voice networks. The cost of base stations is the most important component of large-scale broadband wireless access networks. Such stations are usually placed on towers or on buildings, as is commonly seen for mobile voice networks. These base stations provide the link between the customer and the network.

The number of base stations that is required is a function of two factors—the area that is to be covered and the number of subscribers that are in that area. A minimum number of base stations is required to provide coverage over a given area. This number depends on the physical terrain, the density and height of buildings, and whether a network is designed to provide indoor or only outdoor coverage. This last factor is particularly important. Base stations that support indoor coverage typically have shorter ranges than those for outdoor coverage and may require additional cell-site equipment (such as diversity antennas) to improve reception. Thus, indoor coverage of a given area requires more base stations than outdoor coverage. In addition, these base stations may be more expensive, but they will provide a better standard of coverage and require simpler equipment on customer premises.

The question of indoor versus outdoor coverage can have a major impact on network rollout costs. In a radio-frequency-planning simulation for Addis Ababa, conducted by Alvarion, a manufacturer of broadband base-station equipment, as part of this analysis, it was found that indoor coverage of the target area would require 62 base sites, compared with only eight for outdoor coverage. The base assumption used in this modeling was that, in rural areas, base stations use a lower frequency band of 450 MHz for a range radius of 40 km, while in urban areas, they use the 3.5 GHz band and have a range of 5 km. This is then adjusted according to the type of terrain in which the network is situated.

The second key factor in deciding the number of base stations required is the number of subscribers. In low-density rural areas, the minimum number of base stations required to cover an area is usually sufficient, but in urban areas, where the subscriber density is higher, the number of base stations needed may be well above the basic coverage levels.

Source: Mayer and others 2009.

table A4.3 contains more detailed information on network modeling assumptions.

Data on current broadband coverage in Africa are scarcer than on mobile voice networks, so it is more difficult to define which areas of the country already have access to broadband. The spatial universal broadband analysis was therefore conducted for the whole of a given country's territory, both urban and rural. This approach is likely to overestimate the size of the efficient market gap but should produce a realistic estimate of the split between commercially viable areas and the coverage gap within the model's limits of accuracy.

Two demand scenarios are modeled. The first scenario, referred to here as the *shared-access scenario,* is based on the current patterns of broadband usage found in the Sub-Saharan African countries, where wireless broadband has already been introduced. In these countries, the current customer base is a mix of businesses, high-income residential users, and public access points such as Internet cafés. In this scenario, we estimate the infrastructure that would be required is one broadband subscriber line per 100 inhabitants in urban areas and one broadband subscriber line per 400 inhabitants in rural areas.

Revenues in the shared-access scenario are calculated as 1 percent of GDP in each geographical unit, using the same weighting method for urban and rural GDP per capita as employed in the voice infrastructure model. The cost of the infrastructure is then compared with the revenues that it is likely to generate. This tells us for which areas of each country broadband is commercially viable and therefore what proportion of the population can expect to be living within range of a broadband wireless signal.

Although this shared-access scenario is based on what is seen in the field in Africa at the present time, it represents a level of Internet access that is far below that of high-income countries. In these countries, personalized high-speed broadband access, either to the home or through personal wireless access devices, is emerging as the norm. Operators are developing high-speed networks to provide this level of personal broadband access to customers, and, globally, the market is moving away from shared access in public Internet facilities.

For an understanding of the implications for such levels of provision in Africa, a scenario was modeled based on higher levels of broadband penetration. This scenario is referred to here as the *individual-access scenario.* It assumes a wireless broadband architecture with significantly higher penetration rates: 20 percent penetration in urban areas and 10 percent penetration in rural areas. Revenues in the individual-access scenario are calculated according to assumptions regarding monthly average revenue per user (ARPU). A range of ARPU from $5 to $10 per month was modeled based on trends in voice markets. The key assumptions in this scenario are given in appendix table A4.3.

Results

In the shared-access scenario—based on a mix of business, high-income households, and public Internet-access points—wireless broadband coverage is commercially viable for 75 percent of the population (appendix table A4.4). This means that, operating in a competitive environment, wireless broadband infrastructure can be expected to expand so that about 583 million people in Sub-Saharan Africa would be living within range of a wireless broadband signal. Although people would be able to access the Internet through a range of wireless customer devices, it is expected that, under this scenario, the main form of access would be Internet cafés or other public Internet-access facilities.

Building out broadband network infrastructure to cover the remaining areas would not be commercially viable for the foreseeable future. If network infrastructure is to be developed in these areas, it would require a total subsidy of up to $755 million per year, ranging from nothing in Mauritius and Swaziland to $199 million in the Democratic Republic of Congo. Figure 5.8 presents the results of the analysis by country. These results are dependent on some key input assumptions. The results of varying those assumptions are given in appendix table A4.4.

In the shared-access scenario, differences in demand and costs account for the cross-country variation in broadband's commercial viability. For example, low-income countries generally require higher subsidies because their revenue potential is lower than that of higher-income countries. Cost patterns for wireless broadband are similar to those for basic voice services: Infrastructure requirements are greater in sparsely populated or mountainous countries, and the infrastructure is more expensive, resulting in greater subsidy requirements than in countries with high-population densities.

The results of the individual-access scenario model showed that broadband wireless infrastructure at higher levels of demand (that is, a broadband penetration rate of 20 percent in urban areas and 10 percent in rural areas) is not financially viable anywhere in Sub-Saharan Africa at ARPU levels of $5 to $10 per month. The higher cost of this infrastructure combined with the low willingness of customers to pay means that such a

Figure 5.8 Commercially Viable Broadband Network Coverage (Scenario 1, Shared Access)

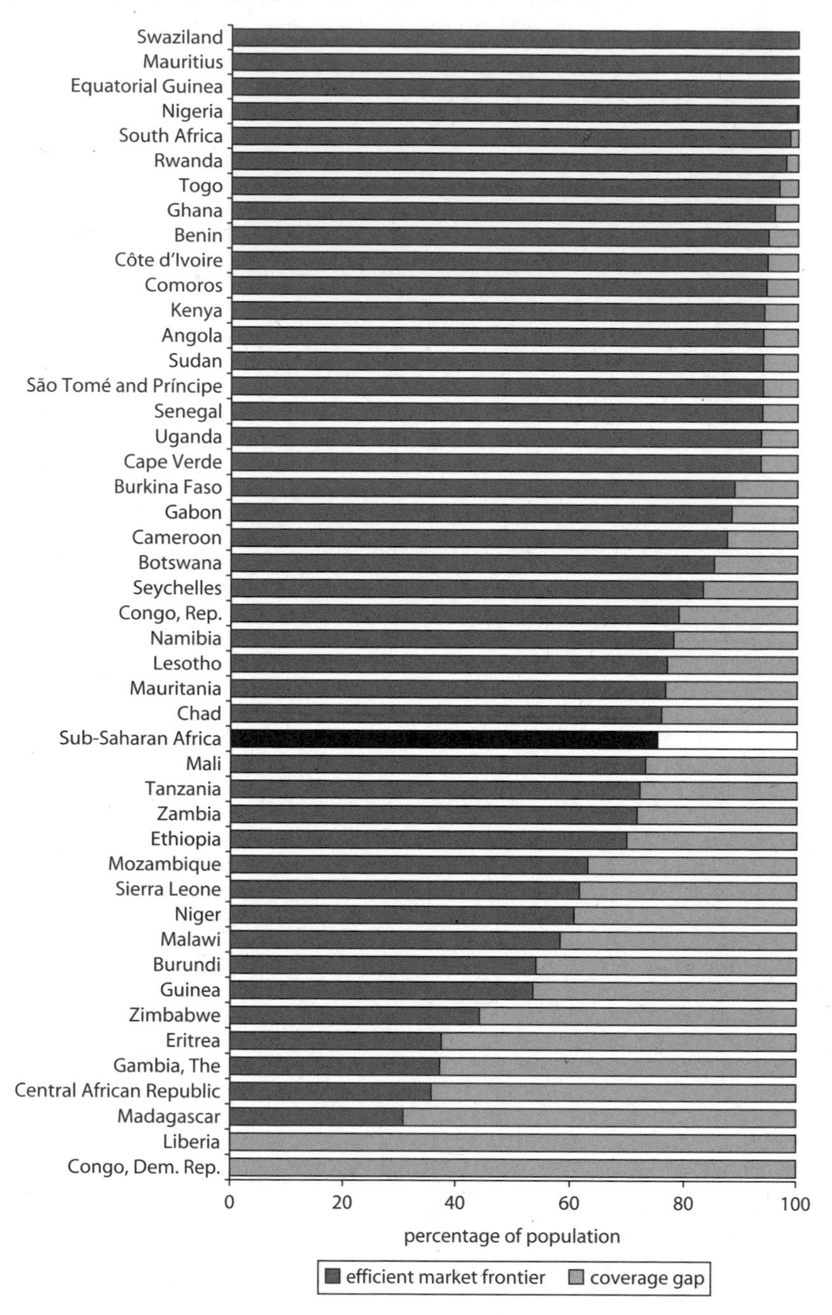

Source: Mayer and others 2009.

network would require a subsidy for it to be sustained. The amount of this subsidy would partly depend on the exact amount of revenue generated. If subscribers spend, on average, $10 per month on broadband services, the total subsidy required to sustain this level of penetration would be $10.5 billion per year. If, however, the average amount spent is $5 per month, which is the level at which many mobile markets in Africa are currently operating (Pyramid Research 2010), the total subsidy required would be $13.3 billion per year (figure 5.9).

The results of both scenarios highlight the importance of international bandwidth costs to the commercial viability of broadband Internet services in Sub-Saharan Africa. A reduction of international bandwidth costs from $2,000 to $400 per month for a 2 megabits per second (Mbps) connection would increase the level of commercially viable coverage in the shared-access scenario from 75 percent to 83 percent of the population (figure 5.10). It would shift about 62 million inhabitants into the efficient market category and would decrease the cost of closing the coverage gap by nearly $200 million per year. The effect of international connectivity

Figure 5.9 Subsidy Requirements for Scenario 2, Individual Access (Selected Countries)

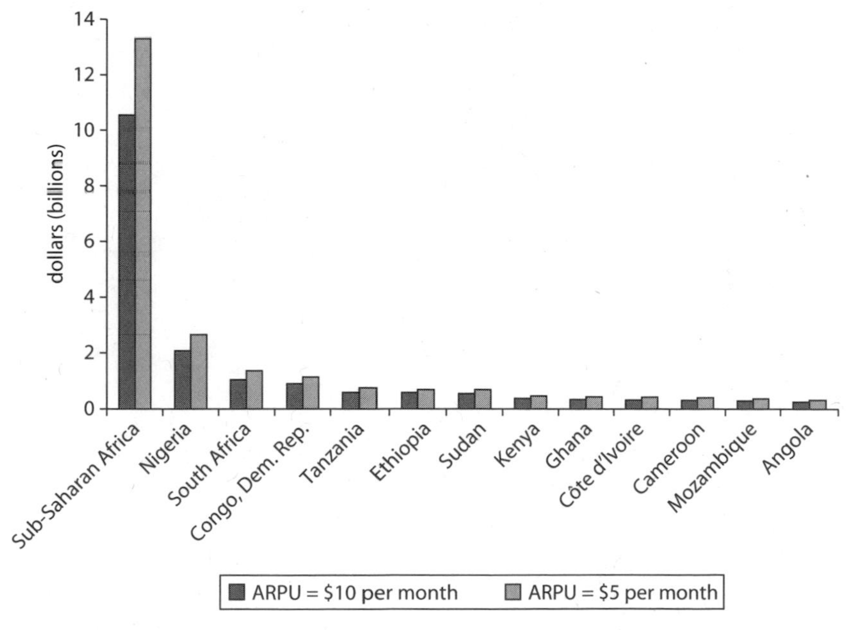

Source: Mayer and others 2009.
Note: ARPU = average revenue per user ($/month).

Figure 5.10 Effects of International Connectivity Costs on Broadband Viability in Scenario 1, Shared Access

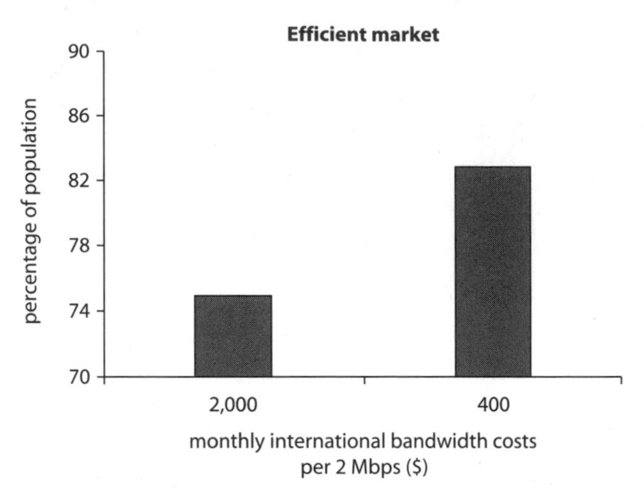

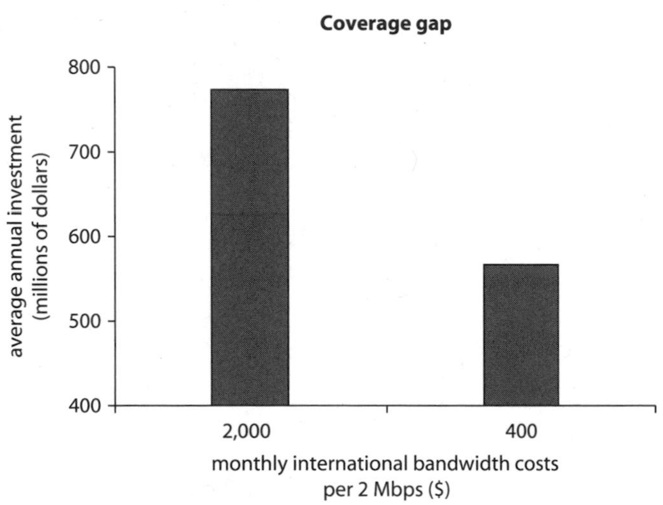

Source: Ampah and others 2009, updated.

costs on the individual-access scenario is as, if not more, striking. The average annual subsidy required to meet the individual-access scenario falls from $10.5 billion to $2.9 billion per year when international bandwidth costs are reduced from $2,000 to $400 per month per 2 Mbps connection.[5]

These results show the importance of submarine fiber-optic networks for the viability of broadband infrastructure in Africa. The increase in international bandwidth combined with lower prices that can be delivered over these submarine fiber-optic networks will dramatically improve the financial viability of the infrastructure and therefore reduce the amount of subsidy that is required to reach universal coverage targets.

These results are as striking as those for mobile voice networks. Broadband Internet in the past has been considered a luxury not commercially viable in Africa. This analysis shows, however, that basic wireless broadband is commercially viable in large parts of the region and that networks could expand to cover nearly three-quarters of Sub-Saharan Africa's population—if private investment is encouraged and effective competition is established. Although no country in the region is close to this level at the time this was written, the experience of wireless broadband in countries such as Kenya and Nigeria gives an indication of the potential demand for the service throughout the region.

The basic broadband being discussed would, however, provide Africa's population with a level of Internet access no longer considered acceptable by customers in high-income countries. Globally, expectations of the quality of the broadband experience have raced ahead as investment has poured into network infrastructure. The infrastructure required to deliver a comparable broadband experience in Africa does not break even at the consumer spending levels considered affordable for the mass market in this study. This does not mean that no feasible market exists for individual broadband service in Africa, but that, unlike mobile voice services, it is not yet widely affordable. The situation may well change over time as wireless broadband technologies and applications mature and become less expensive on a global basis.

The challenge facing African governments is whether to allocate their resources toward subsidies of broadband infrastructure and, if so, how to target those subsidies for maximum effectiveness. Broadband Internet access is simply a pipe, unlike voice telephony, that represents a well-known application for which market demand has been amply proven. Providing the infrastructure is only the first step toward realizing the potential benefits from broadband access. Without concomitant investment in education, technical skills, or facilitated access, publicly funded subsidies of broadband infrastructure could remain stranded and unproductive. The models in this study suggest that shared broadband access underpinned by private sector infrastructure may continue to offer a valuable service to African populations for the medium term as markets expand toward commercially viable levels of individual access.

Notes

1. The analysis in this chapter was originally carried out in 2007 by Mayer and others (2009), using data from 2006. Some parts of the analysis were updated in 2009.
2. This analysis includes only GSM networks. Similar data on networks using the main alternative standard, CDMA, are not available. This does not substantially affect the conclusions of the analysis, because CDMA holds a small proportion of the total mobile voice market in Africa, with only 4 percent of the total voice subscriber market. Angola is one country where there is an extensive CDMA network, but few others have CDMA networks that cover areas of a country in which GSM networks are not present.
3. For countries for which data are available, which varies from year to year.
4. World Bank 2009; World Bank staff analysis.
5. Assuming ARPU of $10/month.

References

Ampah, Mavis, Daniel Camos, Cecilia Briceño-Garmendia, Michael Minges, Maria Shkaratan, and Mark Williams. 2009. "Information and Communications Technology in Sub-Saharan Africa: A Sector Review." AICD Background Paper 10, African Infrastructure Country Diagnostic, World Bank, Washington, DC.

Banerjee, S., A. Diallo, V. Foster, T. Pushak, C. Tsimpo, H. Uddin, and Q.Wodon. 2007. "Access and Affordability of Modern Infrastructure Services: Evidence from Africa." AICD Background Paper 2, African Infrastructure Country Diagnostic, World Bank, Washington, DC.

CIESIN (Center for International Earth Science Information Network, Columbia University), IFPRI (International Food Policy Research Institute), World Bank, and CIAT (Centro Internacional de Agricultura Tropical). 2004. *Global Rural-Urban Mapping Project (GRUMP): Urban/Rural Population Grids.* Palisades, NY: CIESIN, Columbia University. http://sedac.ciesin.columbia.edu/gpw.

Forestier, Emmanuel, Jeremy Grace, and Charles Kenny. 2002. "Can Information and Telecommunications Technologies be Pro-Poor?" *Telecommunications Policy* 26: 623–46.

Gillwald, Alison, ed. 2005. *Towards An African e-Index: Household and Individual ICT Access and Usage across 10 African Countries*. South Africa: Research ICT Africa, the LINK Centre, Wits University School of Public and Development Management. http://www.researchictafrica.net.

GSMA (GSM Association). 2010a. *Taxation and the Growth of Mobile Services in Sub-Saharan Africa.* London: GSMA.

———. 2010b. Untitled data file on GSM mobile network location provided by GSMA. GSMA, London.

Ho, Kelvin. 2005. "Bridging the Digital Divide: Benefits of Using Lower Frequency Bands for Broadband Wireless Access in Remote Areas." Presentation at CDMA Development Group's "CDMA2000 as a Broadband Access Technology" workshop, Hong Kong SAR, China, November 15. http://cdg.org/news/events/CDMASeminar/051115_EVDOWorkshop.

Intelecon. 2005. *Final Report on Expanded National Demand Study for Universal Access Project Part 1–Household Survey.* Report submitted to Nigerian Communications Commission (NCC) by Intelecon Research and Consultancy. Abuja: NCC.

Mayer, Rebecca, Ken Figueredo, Mike Jensen, Tim Kelly, Richard Green, and Alvaro Federico Barra. 2009. "Connecting the Continent: Costing the Needs for Spending on ICT Infrastructure in Africa." Background Paper 3, Africa Infrastructure Country Diagnostic, World Bank, Washington, DC.

Milne, Claire. 2006. "Improving Affordability of Telecommunications: Cross-Fertilisation between the Developed and the Developing World." Paper presented at Telecommunications Policy Research Conference, George Mason University, Fairfax, VA.

Moonesinghe, A., H. de Silva, N. Silva, and A. Abeysuriya. 2006. *Telecom Use on a Shoestring: Expenditure and Perceptions of Costs amongst the Financially Constrained.* Version 2.2 (prepublication). Copenhagen: World Dialogue on Regulation for Network Economies.

Navas-Sabater, J., A. Dymond, and N. Juntunen. 2002. "Telecommunications and Information Services for the Poor: Toward a Strategy for Universal Access." Discussion Paper 432, World Bank, Washington, DC. http://iris37.worldbank.org/domdoc/PRD/Other/PRDDContainer.nsf/WB_ViewAttachments?ReadForm&ID=85256D2400766CC78525709E00616169&.

Pyramid Research. 2010. "Prepaid Mobile Services: Using New Business Models to Boost Profits." Pyramid Research, Cambridge, MA.

Souter, D., N. Scott, C. Garforth, R. Jain, O. Mascarenhas, and K. McKemey. 2005. *The Economic Impact of Telecommunications on Rural Livelihoods and Poverty Reduction: A Study of Rural Communities in India (Gujarat), Mozambique and Tanzania.* London: Commonwealth Telecommunications Organization for Department for International Development.

Stern, Peter, David Townsend, and José Monedero. 2006. *New Models for Universal Access in Latin America.* Montreal/Boston/Madrid: Regulatel/World Bank (PPIAF and GPOBA)/ECLAC Project on Universal Access for Telecommunications in Latin America. http://www.regulatel.org/miembros/ppiaf2.htm.

Ureta, Sebastian. 2005. *Variations on Expenditure in Communications in Developing Countries: A Synthesis of the Evidence from Albania, Mexico, Nepal and South Africa (2000–2003)*. Copenhagen: World Dialogue on Regulation for Network Economies.

Wellenius, Björn. 2002. "Closing the Gap in Access to Rural Communication: Chile, 1995–2002." Discussion Paper 430, World Bank, Washington, DC.

World Bank. 2006. *World Development Indicators.* Washington, DC: World Bank.

———. 2009. *Information and Communications for Development*. Washington, DC: World Bank.

CHAPTER 6

Policy Analysis and Conclusions

Previous chapters have given an overview of the performance of the telecommunications sector in Africa, the development of its infrastructure, and the institutional and market reforms that have driven its growth. They have discussed how this expansion has been financed and attempted to forecast how far current growth rates will take the sector. They have also looked at underlying mechanisms driving the development process and analyzed why some countries have performed better than others and how performance could be improved across the region.

In this concluding chapter, the analysis is taken further to include evolving sector policy priorities, the performance of the region's telecommunications markets, and the drivers of change across the sector. This provides the foundation for a set of recommendations on how sector performance can be further improved and how policy makers can adapt to the new challenges of broadband provision.

Policy Analysis

The design of policy to further boost the development of the information and communication technology (ICT) sector in Africa should begin with an understanding of what the policy objectives for the sector are and what has worked and what has not worked to date.

Sector Objectives

By the end of the 1990s, high-income countries had already experienced the first major wave of the global telecommunications boom. Mobile networks had revolutionized the way that people accessed telecommunications services, and ICT had become integrated into people's lives. The end of the 1990s also saw the beginning of the mobile revolution in Africa. The virtual absence of any competition from fixed-line operators and the falling international costs of mobile network equipment meant that investors could generate high levels of profit from mobile businesses in the region. Throughout the next decade, the telecommunications revolution would spread across Africa, bringing services within reach of many people for the first time.

The overriding policy objective for the sector during this period was to increase the coverage of networks, to reduce prices, and to make ICT as accessible as possible, in particular to the poor and people living in rural areas. Although great success can be celebrated as coverage and access have increased across the continent, the process is far from complete. One-third of Africa's population still does not live within range of a mobile network, and signs suggest that the annual increases in coverage that have been seen for a decade are starting to slow.

Before Africa's ICT boom began, not being able to access mobile services may not have been considered a major handicap for a rural area. As citizens have rapidly incorporated mobile phones into their everyday lives, however, living beyond the range of the networks is increasingly viewed as a social and economic disadvantage. The expansion of networks into rural areas as quickly as possible is therefore even more critical than it was in the past.

Even as mobile networks have expanded and basic voice telecommunications have become a part of everyday life in Africa, the global policy agenda has also changed. In many high-income countries, broadband Internet, once considered a luxury amenity, is now central to people's lives. Businesses use it to improve efficiency and provide direct access to customers while households enjoy fast access to the World Wide Web and all the information and services it provides. Now broadband is seen not only as a telecommunications product but also as a conduit for media content that was previously the preserve of broadcast networks. This global shift in focus away from voice services and toward broadband is also affecting Africa. African countries that fail to develop affordable broadband Internet services are likely to be at an economic disadvantage in the years to come as the world becomes increasingly interconnected.

The two most important broad policy objectives for the ICT sector in Africa are therefore (1) to expand network coverage to all rural areas and (2) to make affordable broadband Internet available to all. Any analysis of performance or forecasts about where the market is going must be done in the context of these objectives.

Sector Performance

One cannot doubt the impressive performance of the telecommunications sector in Africa since the end of the 1990s—coverage rates and subscriber numbers speak for themselves—but the success of mobile services at the aggregate level can hide less successful aspects of the sector. The number of fixed-line subscribers is growing only slowly and is, in many countries, declining. Mobile networks have expanded into rural areas, but about half of Africa's rural population still does not live within range of mobile services. Meanwhile, some countries are far behind even this benchmark. Although Internet-access rates are growing across Sub-Saharan Africa, they remain well behind comparable regions of the developing world. As an increasingly interconnected world embraces the Internet, Africa's low rates of access will become more problematic.

Since the beginning of the market liberalization process in Sub-Saharan Africa, prices have remained high. This has had the benefit of making the sector very profitable for most operators, which has attracted investment and stimulated competition, but high prices also limit access, in particular for the poor. Prices for mobile services have now begun to fall as competition intensifies. As markets have begun to mature, operators are switching their competitive strategy from focusing primarily on network coverage to focusing on price. This is beginning to drive down prices—but these still remain well above those in comparator regions, such as South Asia. The price of Internet access remains prohibitively high in the majority of countries in Sub-Saharan Africa, and the Internet is more expensive here than in other parts of the world.

Improving the quality of ICT services is a challenge to governments and regulators throughout Africa. Although limited quantitative data on service quality are available, the everyday frustrations of users are well known: difficulties in establishing connections to other subscribers, particularly at peak times of the day, and calls being dropped by the network midway through a conversation. These problems typically get worse as operators overstretch their networks by carrying more traffic than the networks can handle. The situation can be improved if regulators introduce quality standards and then effectively monitor their application, but this has yet to be done systematically on a regionwide basis.

Broadband infrastructure in the region remains limited but is showing signs of improving in the best-performing countries. Operators and Internet service providers (ISPs) are developing wireless broadband access networks with technology based on global wireless broadband standards and are upgrading backbone networks using fiber-optic cable technologies capable of carrying large amounts of traffic at low costs. The customer response is encouraging. In Nigeria, for example, the number of broadband wireless subscribers has reached 10 million since the launch of the services in 2006—nearly twice as many as in South Africa. Although the region's submarine fiber-optic cable infrastructure has been late to develop, the recent surge in investment in cables has already resulted in rapid increases in Africa's total bandwidth and a drop in prices—developments that are expected to continue over the next few years as new cables come on line.

The Drivers of Change

The successful development of the voice market in Africa has been driven by market liberalization, the establishment of competition, and effective regulation. Relevant policies have been adopted in most countries, prompting a positive market response: investment in network expansion, innovation in service provision, and, eventually, reductions in prices. Variations in the extent to which these drivers have been implemented go a long way in explaining the differences in country performance across the region, although incomes and geography are also important factors. Completing this reform agenda is therefore essential for the further development of the sector.

The first stage in the reform process is market liberalization, which is usually implemented by issuing new licenses, typically for mobile network operators, ISPs, and, eventually, fixed network operators. Almost all countries have implemented such market reforms to some extent, but the degree of liberalization varies across the region. The leading reformers have shown that it is possible to have many more players in the market than originally was thought. Further liberalization is key to the future development of the sector.

The second driver of reform, which goes hand in hand with market liberalization, is the establishment of regulatory frameworks and institutions. This requires legal reforms that establish a sound, transparent framework for investment and competition. Such a framework would necessarily also include regulatory institutions to oversee the market. These institutions should be legally empowered and politically independent so that they can carry out their functions effectively. A high level of institutional capacity is also needed to ensure that regulatory decisions are made on a sound technical basis.

The process of market liberalization introduces private investment and market competition to the sector. Although it is possible for one of the players in the market to be owned by the state, this has adverse long-term consequences for sector development. State ownership of one of the telecommunications players in a liberalized market creates incentives for governments to skew regulation and policy in favor of the state-owned enterprise over others. This has an adverse effect on competition—the ultimate driver of development in the sector. Privatization of state-owned enterprises is therefore an important driver of long-term development in the sector.

Broadband presents new challenges for policy makers around the world, including Africa. The model for reform of the broadband market in Africa is not yet as well established as it is for voice, but lessons are quickly emerging from the leading markets in the region. One key lesson is that it is important to appreciate the full broadband value chain and to understand the drivers of reform at each stage. It is important to consider issues such as international connectivity, domestic backbone networks, access networks, customer equipment, and service and pricing. The same principles of investment and competition apply to all these factors, but the measures needed to implement them are different.

Africa's lack of international bandwidth is being addressed by high levels of investment in submarine cables. Ongoing success depends on downstream users being able to access submarine-cable-landing stations at low cost and on nondiscriminatory terms. This is best achieved through competition between cable-landing stations, but, where this is not possible, regulatory authorities will need to step in to provide regulated access.

The absence of extensive backbone networks in Africa could present a significant obstacle to its broadband market. For broadband to be a viable, affordable option for the masses, networks must utilize fiber-optic technology. Private investors are investing heavily in fiber networks, and competition is evolving, but only along major trunk routes. Expanding the reach of fiber-optic networks to small towns and rural areas requires government support. Policy makers need to boost investment in these areas without undue interference in the market.

The cost of wireless broadband equipment is dropping as international competition in manufacturing intensifies. Leading telecommunications markets in Africa are already upgrading their networks to third generation (3G) and other types of broadband wireless access. This process will continue: It is expected that much of Africa's mobile network infrastructure will become broadband enabled over the next 10 years. The cost of the equipment that customers use to access broadband remains a barrier, however, particularly in low-income countries. The cost of such customer premises equipment (CPE) varies according to the technology used—fixed-wireless CPE is typically more expensive than broadband-enabled mobile handsets—but the cost of all types of CPE is falling. International competition in the manufacture of these customer devices and increasing integration between broadband-access equipment and computers will further push down the cost of accessing broadband. Rapid recent innovations in the market for smart phones, netbooks, and tablet computers will continue to make it easier for people to gain access.

Customer service has been a major barrier to broadband market development in Africa. The key customer service innovation in the voice market in Africa was the introduction of prepaid mobile services, which dramatically reduced the commercial risks facing network operators and made it easier for customers to control expenditures. Broadband in high-income countries has developed on a subscription (that is, postpaid) model in most countries, but this is unlikely to be suitable for Africa for the same reasons that postpaid mobile voice services were not. Prepayment systems are available for broadband, particularly wireless broadband, and their adaptation to the African market will be key to the success of broadband in the region.

Recommendations

Recommendations fall into two broad groups: completing the reform agenda and creating incentives for operators to meet evolving policy objectives.

Completing the Reform Agenda

The reform process that has driven the improvements in the ICT sector in Africa is not yet complete. Some countries lag far behind others, and the region as a whole lags behind other leading developing countries. Completing this reform agenda should therefore be a major strategic objective for the sector. This agenda can be divided into two parts, both aimed at promoting effective competition in the sector: full liberalization and effective regulation.

Liberalization. Effective market liberalization requires more than just issuing one or two mobile licenses. The liberalization process needs to go further and deeper through issuing more licenses and reforming the licensing framework itself.

Issue more licenses across all segments of the market. Entry into the telecommunications market in every country in Africa continues to be controlled through the licensing process. By requiring companies to obtain a license before investing in the sector, governments control how many players are in the sector—and who they are. The license-award process is also an important means of raising substantial amounts of revenue in the form of license-award payments and regular fees. The global trend is moving away from this model of tight control over market entry. The European Union, for example, has replaced its licensing regime with an "authorization" regime that has significantly reduced the discretion of public authorities over which companies enter the sector (Flanagan 2009). Some African countries have reduced the conditions for getting certain types of licenses (for example, for ISPs), a step that goes far toward liberalizing any given segment of the market. For several parts of the market, however—in particular, the building and operating of communications network infrastructure—most countries retain tight controls through the licensing process. Apart from managing the use of radio spectrum, little economic rationale can be seen for such controls. The experience of some very small African countries, such as Burundi (which has five active mobile operators and one additional licensee that is not yet fully operational), shows that markets can sustain many operators.

Reform the licensing framework. Under the typical telecommunications licensing system in Africa, the licensee is permitted to carry out a tightly defined set of activities such as network construction, mobile voice service provision, and Internet service provision. Any existing licensee wishing to invest in providing a new service or new type of infrastructure usually has to go back to the regulatory authority to obtain additional permission. The impact of this system is to slow investment, limit innovation, and restrict competition. These regulatory controls on the activities of licensees also serve, either explicitly or implicitly, to protect certain operators—usually the state-owned incumbent operator. A good example of this is the restriction on network operators from providing backbone services for third parties (that is, on a wholesale basis to other operators and ISPs). Such restrictions have often been placed on operators to protect the incumbent operator—which, at the outset of the liberalization process, usually has the only fiber-based backbone network. The effect of such restrictions is to limit investment and competition because it is reducing the potential market for an operator considering investing in such infrastructure. The licensing framework should be reformed to reduce and preferably eliminate restrictions on the activities of operators, subject to controls on the use of radio spectrum. It should also take account of technological convergence, a global trend that is breaking down the traditional one-to-one relationship between networks and services. This will increase investment, promote competition, and stimulate innovation in technology and services. In 2005, for example, Tanzania introduced a converged licensing framework with four license categories: network facilities, network services, applications services, and content services. As of June 2010, the regulator had issued 16 national and 8 international network facilities licenses, which allow licensees to offer any facilities-based telecommunications service. By following this type of regulatory reform, other countries will also encourage investment competition and innovation in the ICT sector.

Avoid reintroducing restrictions on competition. The recent trend toward tighter management of international gateways (to collectively raise international termination charges) is a reversal of the process of liberalization that has been so successful in Africa. It is likely to have an adverse impact on sector development and could reduce sector tax revenues in the long run. It will also have a negative impact on telephone users because they will receive fewer international calls, and this makes a country less attractive to foreign investors. Such measures should not be introduced and, where they exist, should be removed.

Privatize telecommunications operators that currently remain under state ownership. State ownership of telecommunications operators provides few benefits to a country. Such operators frequently have a small market share, are often inefficiently run, and are usually subsidized by the state, either explicitly through favorable tax and license-fee treatment, or implicitly through regulatory rules skewed in their favor. Despite this protection (or perhaps because of it), state-owned operators have generally performed poorly and, in most cases, have failed to compete successfully with privately owned companies. The long-run cost of state ownership is that investment and competition are constrained by policies and regulations designed to protect these operators. Half of the countries in Sub-Saharan Africa have privatized their incumbent operators. The other half continue to maintain them under state ownership. This ownership structure makes it difficult for these operators to compete, and so, as the sector expands and competition intensifies, the value of state-owned companies frequently declines and becomes a greater drain on public resources. These companies should therefore be privatized as quickly as possible to obtain the maximum sale value and to boost the long-term growth of the sector.

Regulation. Effective regulation is at the heart of the reform agenda. Regulators play a central role in ensuring that sector policy is implemented and that competition develops effectively. The effectiveness of sector regulation is therefore a key part of the reform agenda. Many different aspects are part of improving regulatory performance.

Ensure that regulators are independent of government. The mandate of regulatory authorities is to implement government sector policy. This requires technical decisions without undue political influence and without favoring particular industry stakeholders. The institutional independence of the authority is one of the factors that influence how well it can carry out this mandate. In practice, two aspects of regulators' institutional design are particularly important: how senior management is appointed and how institutions are financed. Senior management, especially the head of a given institution, should be protected from undue political influence, in particular from the sector ministry. Ideally, management should be appointed by a higher political authority such as the cabinet, the parliament, or the head of state. Terms should be fixed and protected—except in the case of corruption or failure to perform duties—so that they are shielded from short-term political pressures. The way in which an institution is financed also plays a role in determining its level of independence. In most cases, the regulator is financed from sector levies, but the budgeting is controlled to a greater or lesser degree by the government or by the legislature. A balance needs to be struck between (1) ensuring that the funding of the institution cannot be used to exert political pressure on regulatory decisions and (2) establishing some form of oversight by government institutions. Countries should take steps to increase the independence of their regulatory institutions and thus to improve the quality of regulatory decisions.

Strengthen the legal powers of regulators to implement regulatory decisions. Regulatory decisions often meet with opposition from stakeholders within the industry, in particular large or politically well-connected operators. This can sometimes result in these stakeholders taking legal action to prevent regulators from implementing decisions. Such action is often seen when regulators attempt to reduce interconnection charges because this has a direct impact on operator revenues, even though it is to the benefit of customers. An important way of improving the effectiveness of regulatory authorities is by ensuring that they are given sufficient legal powers to design and implement decisions, after an appropriate investigation and consultation process. Regulatory authorities also need adequate professional staff with a legal background to be able to exercise their regulatory powers effectively.

Improve the regulation of interconnection. The terms by which networks interconnect and then carry traffic that originates on different networks are a key determinant of the price and the quality of service that customers receive. Left to negotiate themselves, operators typically push the prices that they charge one another for termination well above cost, and these higher wholesale prices feed through into higher retail prices. Operators also often underprovide interconnection points so that subscribers experience difficulty in connecting to a subscriber of another network. These interconnection arrangements require tight regulation, and the global trend has been to push interconnection charges down toward cost through regulatory controls. The calculation of network costs of termination is a complex and expensive process, and the results are often met with resistance from the operators. Yet as more attention is paid to the price of calls and the quality of service, African regulators need to focus their efforts on this area. Fair, cost-based regulatory controls of interconnection would go a long way toward stimulating competition and lowering prices. Such measures require regulators to invest in capacity building and to engage high-level external advisory services where appropriate.

Improve the allocation and management of radio spectrum. Radio spectrum is a crucial resource of the telecommunications sector in all countries, and the lack of alternate wireline networks in Africa means that it is particularly important there. The allocation and management of the spectrum can have a major impact on the value of operators' investments, and arbitrary changes in spectrum allocations can have a significant disruptive effect on the market. Spectrum allocation and management practices in Africa are often a disincentive to market growth and innovation. Governments and regulatory authorities should consider the global trend toward predictable, transparent, and more market-based mechanisms for spectrum allocation and management. This will improve investor confidence in the sector and allow operators and ISPs to innovate and bring in new technologies to improve the services provided to the public.

Introduce other procompetitive regulatory measures. Full and effective competition requires a suite of proactive regulatory measures following the establishment of the basic foundations of a liberalized market structure. International experience provides many examples of such measures, including mobile number portability, mobile virtual network operators, collection, and publishing price and quality-of-service data. African regulators should consider each of these measures and determine whether they could be applied in their jurisdictions to promote competition. As competition develops, the primary role of the regulatory authority evolves from controlling the market power of a dominant operator to regulating competition. The legal framework within which the regulator operates and the technical capacity of its staff should change to reflect this evolution in the market.

Promote facilities sharing through proactive regulatory initiatives. Facilities sharing has the potential to significantly reduce costs for operators in the market. In some cases it is established through market forces, particularly through the outsourcing of mobile tower infrastructure, which is then shared across multiple operators. Regulators can help the commercial process by ensuring that appropriate regulations are in place and that any commercial arrangements between tower companies and operators do not adversely impact competition through, for example, agreements on excluding competitors from access. Cases are also found in which commercial arrangements for facilities sharing are not emerging: The parallel development of fiber-optic cable infrastructure is one example. In these cases, regulators should consider a more assertive form of facilitation through the preparation of model agreements for sharing infrastructure or by obligating operators to share where technically feasible. Finally, cases are seen where the regulator needs to impose sharing on select operators, particularly in cases where they are dominant in a segment of the market or control an essential facility such as a submarine fiber-optic cable-landing station. This is a difficult task and should be undertaken only where competition has failed to make an impact on the operator's market position. In cases such as submarine-cable-landing stations, it can be essential for the successful development of competitive markets. African regulators could also follow the examples of other regulatory authorities that have imposed similar rules (Telecommunications Regulatory Authority [Bahrain] 2008).

Strengthen the human capacity of regulatory authorities. To a considerable extent, the effectiveness of a regulatory authority is determined by the skills, knowledge, and experience of its staff. Developing this human capacity should therefore be a top priority for regulators. Regulators should invest in training, the provision of technical advice, and capacity building on an ongoing basis to ensure that their staff maintain high levels of skills, particularly in areas of new technology and global trends in regulation.

Promote regulatory harmonization through regional bodies. Considerable similarities exist across telecommunications markets in Africa. This provides an opportunity for mutual support and learning. The increasingly international nature of telecommunications in the region—stimulated by the launch of multiple submarine fiber-optic cables—means that cross-border regulatory harmonization is becoming more important. Many regional policies and regulatory settings that provide the institutional mechanisms for such initiatives should be further promoted and sustained.

Creating Incentives for Operators to Meet Evolving Policy Objectives

It is clear that, although liberalization and competition meet many of the sector objectives, the sector is unlikely on its own to meet all of them—such as, for example, 100 percent mobile network coverage. An important role of the public sector is therefore to provide incentives for companies to meet these objectives. These incentives can take many different forms. At one end of the range are changes to technical aspects of the regulatory framework; at the other are financial subsidies for companies meeting policy objectives. Some of the key incentives are highlighted and discussed here.

Universal service. Ensuring that affordable ICT services are provided in rural areas is a major policy priority. Regulators should carefully consider how to promote this in the context of liberalized and increasingly competitive markets. Liberalization has boosted investment, expanded network coverage, and lowered prices, thereby meeting many aspects of the universal service policy objective. If effective competition is sustained, very large amounts of private investment will be attracted to the sector over the next 10 years, and coverage is expected to expand to about 90 percent of the region's population. The central strategy for achieving universal access is therefore further deepening competition. Despite this, it is clear that this strategy, on its own, will not achieve 100 percent coverage because some areas—typically remote and rural—are not likely to be commercially viable for the foreseeable future. Regulatory strategy for providing coverage in these areas, in a liberalized market structure, should be based on providing commercial incentives for operators to invest. This can be done in various ways.

Tax policy. Marginal areas of the country can be made more attractive to operators by reducing taxes on equipment and on ICT services rather than, as has often been the case, elevating the level of tax on the sector. Taxes can also be used to give direct incentives to operators providing coverage in rural areas. In exchange for operators investing in these areas that would not otherwise be covered, taxes or sector levies could be reduced. Import duties or other forms of trade protection on inputs for telecommunications operators also have an impact on overall costs. Reducing the trade protection of local component suppliers such as battery manufacturers, for example, would reduce the costs of a component important to the deployment of networks in rural areas.

Emerging technologies. Competition in the global market for GSM equipment has driven down prices dramatically. Innovations in technology—such as low-capacity coverage extenders, use of lower-frequency bands (for example, 450 megahertz), extended coverage base stations, and solar power supplies for base stations—are being deployed in countries with significant rural populations, such as Australia, China, and India. By encouraging the use of these innovations, governments can reduce costs and increase the financial viability of GSM networks covering rural populations in Africa.

Other cost-reduction strategies. Regulatory measures such as infrastructure sharing (discussed above) can be used to reduce costs for operators, particularly in rural areas. Other factors, such as access to power, increase costs and thereby reduce the financial viability of rural areas of the country. Measures such as sharing power facilities or allowing mobile operators to sell power into local or national grids can also help defray the costs of telecommunications network power generation. Another key cost area for telecommunications operators is skilled staff. Restrictions on the movement of such labor push up operating costs and contribute to making rural areas financially unviable.

Revenue-enhancement strategies. On the other side of the equation are the revenues generated by customers using the network. If these can be enhanced in rural areas, some of these strategies could become financially viable. This can be done in several ways. The increased use of ICT to deliver public services has the potential to increase traffic in rural areas. Equipping health workers or agricultural extension workers with phones, for example, raises demand in rural areas. If this could be done in coordination with operators, it could provide sufficient incentive for them to extend their networks. Similar "anchor tenant" approaches can be used by equipping government offices with ICT equipment and services and committing to medium-term service provision contracts. Another approach is to eliminate any regulatory constraints on providing value added services via mobile phone networks; a good example of this is mobile banking. By removing potential regulatory restrictions on providing banking services via mobile phones, the government of Kenya allowed Safaricom to launch its M-Pesa banking service. The use of its mobile network for banking services has increased communications traffic and customer loyalty, thereby increasing the network's profitability. Such schemes are valuable particularly in rural areas, which do not have alternate financial services. By encouraging them through effective regulation, governments are indirectly boosting the likelihood that operators will extend networks into rural areas.

Even in the most favorable policy and incentive environment, in most countries in Africa a small but significant proportion of the population will be found in areas that are not commercially viable. Some form of direct financial subsidy is therefore needed to provide an incentive for operators to expand their networks into these areas. Many governments and regulators have introduced such subsidy schemes through universal service funds. These are funds that are typically financed through levies on the sector and are then, in principle, channeled back to the operators to provide subsidies for rollout into rural areas. The performance of these funds in Africa, as in other parts of the world, has not always been satisfactory. In many cases, revenues from the sector levy are collected but remain unused. Other funds are oriented toward fixed-line rollout in rural areas despite the lower costs of installing wireless systems. Governments have also had difficulty designing mechanisms that do not adversely affect competition in rural areas and are sufficiently flexible to work in a rapidly changing market environment. One example of this is in Uganda, where, despite the universal service program being regarded as an overall success, at the outset, the subsidy scheme struggled to adapt to the rapidly changing telecommunications market. In other countries, lack of effective competition in the market in general has prevented an effective competitive subsidy allocation process from taking place. Around the world, lessons are emerging that should be applied in the design of such systems in Africa.

- *Implement direct-subsidy mechanisms as a "last resort" policy measure.* Competition has proved to be a much more effective way of providing access for the majority of rural areas. Universal service funds financed from sector levies raise overall prices, thereby excluding the poorest from accessing services. The size of these levies should therefore be kept to a minimum. Ideally, direct-subsidy schemes should be implemented only once network growth begins to show signs of slowing down. This would minimize the disruptive effects on the development of competition.

- *Competitive subsidy allocation schemes do not work in uncompetitive markets.* Multiple operators are needed that are capable of providing the services identified in the competitive subsidy allocation process if such a process is to work. In uncompetitive markets, bidders sometimes refuse to participate in the process, or the lack of effective competition results in higher prices for all users and increases the cost of the subsidies required.

- *Direct-subsidy schemes are more effective if they promote competition rather than exclude it.* For example, subsidies targeted at shared infrastructure, such as towers that can be used by multiple operators, are more effective than a subsidy that benefits only one operator. This type of shared-tower-infrastructure-subsidy mechanism has been used effectively in South Asia to extend networks into rural areas but has not yet been extensively used in Africa. A similar approach can be used for backbone infrastructure (see below). Direct-subsidy mechanisms are also more effective if they are combined with other incentives such as tax-based incentives and facilities sharing.

Broadband. Broadband is a new area in the telecommunications sector in Africa. The design of policy to promote it is therefore more challenging than in the mobile sector, for which one finds ample evidence to draw upon. Broadband is also a complex product that requires many different links in the value chain to be delivered effectively. An effective broadband strategy therefore needs to address the entire value chain. Many of the important elements of such a strategy are regulatory in nature. For example, the licensing framework and associated regulations (discussed above) have a major impact on the incentives for operators to invest in broadband technology and infrastructure. A role may also be found for direct financial incentives to stimulate the rollout of broadband network infrastructure and services.

Submarine fiber-optic cables. The overwhelming majority of financing for the submarine fiber-optic cable infrastructure in Africa has come from private investors or telecommunications operators that are state owned but operating as commercial entities. A few exceptions to this can be identified. The Eastern African Submarine Cable System (EASSy) benefited from public sector support through the financing of the preparation and feasibility studies by the World Bank. The East African Marine System (TEAMS), the cable that connects Kenya to the United Arab Emirates, was initially cofinanced by the government of Kenya, which retains a minority stake in the cable. This was done in the early stages of cable development along the east coast. Subsequent cables in this part of the region (SEACOM, Lower Indian Ocean Network [LION]) did not require public financing. More recently, local operators in Liberia, São Tomé and Príncipe, and Sierra Leone did not have sufficient resources to join the consortium developing the Africa Coast to Europe (ACE) submarine cable that runs along the west coast. Public financing was made available to ensure that these countries did not miss out on the opportunity to be connected to the global high-speed communications networks. Other than the specific cases of small countries that do not have sufficient resources, or countries that lie off the main routes of submarine cables, a strong rationale cannot be advanced for direct financial incentives to attract investment in submarine fiber-optic cable infrastructure in Africa. Once the cables currently under construction are complete, Africa will have access to a large amount of international bandwidth, supplied on a competitive basis.

The presence of submarine cables, on their own, however, is not sufficient to guarantee low-cost bandwidth. Monopoly control of the cable-landing stations can also allow operators to maintain high prices and adversely affect competitors. Competition between landing stations is needed to ensure that the low prices of offshore submarine-cable capacity are available to customers. Governments should try to ensure that cables have multiple landing stations owned by different operators in their countries. Where such duplication of landing stations and competition between them is not possible, governments should focus on ensuring that open access to them is available. This is best done by allowing open participation in ownership and operation of the facility by players in the market. Failing that, tight regulation of the landing facility is needed, but it is important to recognize the challenges that this presents. In practice, such regulation has proved difficult to implement effectively.

Wireless access networks. International competition is growing in the manufacture of wireless broadband access network equipment; prices are falling quickly, and this process is likely to continue as operators in high-income countries increasingly invest in wireless broadband networks. In many African countries, much of the mobile wireless network infrastructure is capable of being upgraded to 3G at relatively low cost. The other wireless broadband standard that is currently widespread in Africa is Worldwide Interoperability for Microwave Access (WiMAX), and many WiMAX operating and spectrum licenses have been issued. Looking ahead, Long Term Evolution (LTE) is likely to be a significant global wireless standard that will be adopted by many mobile operators. The barriers to entry in this part of the market are therefore relatively low, and already operators and investors have expressed considerable interest in developing this type of infrastructure. The analysis in this study indicates that competition could push wireless broadband networks to cover a considerable proportion of Africa's population. Therefore, at this stage of market development, one apparently cannot identify any clear rationale for direct public financial support for this segment of the infrastructure, other than as part of regular universal access programs. Regulating access to spectrum for the new standards will, however, be a crucial determinant of the success of wireless broadband in Africa.

Backbone networks. The experience of fiber-optic-backbone network development in Africa to date shows that the private sector is willing and able to invest large amounts in fiber-optic-backbone infrastructure, but this investment will be focused on the most profitable areas, primarily trunk routes connecting major urban areas to one another and to coastal landing stations or land borders. Beyond these areas, it seems likely that a role will be seen for the public financial support for network rollout. Some lessons are already emerging from the experience of such projects in Africa:

- *Target public support in rural areas.* Public support for fiber-optic infrastructure is likely to be needed only outside major urban areas and away from profitable trunk routes. Public support for backbone network infrastructure should therefore be targeted at these areas. If it is targeted at profitable routes, it could crowd out private investment. It may also lead to the same policy and regulatory incentives problems that are seen when the state owns one of the operators.

- *Provide financial support in partnership with the private sector.* Private operators have much experience in building and operating communications networks. They are also the primary users of any networks built with public support. It is therefore essential that such public support be channeled through partnership with these private parties, which can be done in many ways. At a minimum, private operators should be involved in the network design. Where possible, they should also coinvest with the government so that they have a stake in the financial and operational success of the networks.

- *Public support should be procompetitive.* Backbone networks represent a major fixed investment. Investment by one operator can therefore block further investment by other operators to the detriment of competition. Public financial support for backbone network development should encourage competition rather than discourage it. A good example of this is the government of Rwanda's backbone project. The government built a backbone network with additional ducts, which were then leased to private operators. The public investment in the network stimulates further private investment and encourages competition between these private parties. Such models are a good example of how public investment can "crowd in" private investment and support a competitive market.

- *Leverage existing infrastructure.* Governments can also facilitate access to alternative forms of infrastructure such as railways, pipelines, and electricity networks. These networks provide routes along which fiber-optic cables can be quickly laid at relatively low cost. Because they are often also secure networks, they also generally provide a more secure communications infrastructure than fiber cables laid underground. The quickest and simplest way of encouraging this process is to allow the companies that operate these networks to develop their own fiber-optic networks and then provide them with licenses to provide telecommunications services. The Zambia Electricity Supply Corporation Ltd. (ZESCO) and the Electricity Supply Corporation of Malawi (ESCOM) are examples of electricity companies that have played a significant role in the development of fiber-optic communications infrastructure in their respective countries.

- *Encourage infrastructure sharing.* Regulatory authorities can also facilitate infrastructure sharing, which is proving to be an effective way of reducing the cost of fiber-optic network rollout. This is already taking place in numerous countries through commercial agreements between operators. Regulators can ensure that such sharing arrangements do not discourage competition by prohibiting activities such as price fixing and by ensuring that sharing arrangements are made available to all parties.

- *Ensure that complementary regulatory measures are in place.* Efforts to provide public financial incentives to invest in backbone infrastructure without a conducive regulatory framework can be self-defeating. For example, it would be inefficient to provide such incentives while, at the same time, restricting the types of services that these backbone network operators can provide. The first priority when considering financial subsidies for backbone network developments is therefore to remove obstacles to investment, which include limits on the number of licenses and constraints on the services that networks can sell.

- *Use demand-side measures to boost network development.* Governments can use their purchasing power as a buyer of backbone services to reduce the commercial risk to investors by signing prepurchase agreements. This approach was successfully used in the Republic of Korea to stimulate a major rollout of broadband network infrastructure. The major advantage of this approach is that it does not unduly interfere with the competitive market structure in which private parties build, own, and operate infrastructure.

Conclusion

The story of the ICT sector in Africa since the end of the 1990s has been one of success. Policy reform has opened the market to private investment while competition has driven the expansion of networks, the reduction of prices, and the tailoring of services to suit the specific needs of the population. But this process is far from complete. Many countries still maintain restrictions on the activities of operators in the telecommunications sector and, in cases such as the introduction of international gateway operators, are actually reversing the process of market liberalization that has been so successful. The reform agenda is therefore yet to be completed in every African country, and, until it is, many Africans will not be able to access basic telecommunications services at affordable prices.

Yet, just as African countries are making progress with the rollout of mobile services, the global policy agenda is moving on. The pace of broadband take-up around the world is accelerating, and Africa is at risk of being left behind. Countries urgently need to reconsider their sector policy strategies if they are to see broadband take off in Africa as it has in other parts of the world. Competition and private investment will help, but governments may have to play a more assertive role in the sector if they are to see the rollout of the high-capacity network infrastructure their countries need. Important steps are filling infrastructure gaps, helping the private sector to invest and compete, and designing sound and productive policy. If countries get these and other elements right, the reward will be enormous—a dynamic and growing broadband market, giving Africans affordable access to the global communications community. If not, Africa is at risk of missing out on the second ICT revolution that is sweeping the world.

References

Flanagan, Anne. 2009. "Authorization and Licensing." In *Telecommunications Law and Regulation*, 3rd ed. ed. Ian Walden, 295–389. Oxford: Oxford University Press.

Telecommunications Regulatory Authority (Bahrain). 2008. "Telecommunications Infrastructure Deployment: A Guideline Issued by the Telecommunications Regulatory Authority." Telecommunications Regulatory Authority, Manama.

APPENDIX 1

Access to Telecommunications in Africa

Table A1.1 Submarine Fiber-Optic Cables Connecting to Africa, 2010

System	*Type*	*Cost ($ million)*	*Distance (km)*	*Capacity*	*Status*
SAT-1	Consortium	n.a.	9,800	23 Mbps	Exited service
SEA-ME-WE II	Consortium	n.a.	18,000	565 Mbps	Exited service
SEA-ME-WE III	Consortium	n.a.	27,000	20 Gbps	Operational
SAT-2	Consortium	250	9,500	560 Mbps	Operational
Atlantis-2	Consortium	240	8,500	10 Gbps	Operational
SAFE	Consortium	639[a]	13,800	130 Gbps	Operational
SAT-3/WASC	Consortium		14,350	120 Gbps	Operational
SAS-1	Consortium	n.a.	333	1.28 Tbps	Operational
FLAG FALCON	Private	n.a.	10,300	2.56 Tbps	Operational
EASSy	Hybrid consortium	235	9,900	1.4 Tbps	Operational
EIG	Consortium	700	15,000	3.84 Tbps	Operational
SEACOM	Private	650	13,000	1.28 Tbps	Operational
TEAMS	Public-private partnership	110	4,900	320 Gbps	Operational
LION Cable	Consortium	n.a.	1,800	1.3 Tbps	Operational
LION 2 Cable	Consortium	79	3,000	1.3 Tbps	Under construction
GLO-1	Private	170	9,330	318.4 Gbps	Under construction
MAIN-1	Private	240	6,900	1.92 Tbps	Under construction
WACS	Consortium	600	14,000	3.84 Tbps	Under construction
ACE	Consortium	650[b]	14,000	1.92 Tbps	Under construction
Total	Operational	2,800		12.3 Tbps	12
Total	Under construction	1,700		9.0 Tbps	5

Source: Hamilton 2010.

Note: ACE = Africa Coast to Europe; EASSy = Eastern Africa Submarine Cable System; EIG = Europe-India Gateway; FALCON = FLAG Alcatel-Lucent Optical Network; FLAG = Fiber-Optic Link around the Globe; GLO-1 = Globacom-1; LION = Lower Indian Ocean Network; SAFE = South Africa Far East; SAS-1 = Saudi Arabia–Sudan 1; SAT-1 = South Atlantic 1; SAT-3/WASC = South Atlantic 3/West Africa Submarine Cable; SEA-ME-WE = South East Asia–Middle East–West Europe; TEAMS = The East African Marine System; WACS = West Africa Cable System; Gbps = gigabits per second; Mbps = megabits per second; Tbps = terabits per second; n.a. = not applicable.

a. The combined value of the SAT-3/WASC/SAFE cable system.

b. Estimated.

Reference

Hamilton, Paul. 2010. "Broadband Network Development in Sub-Saharan Africa." Unpublished paper. Hamilton Research, Bath, England.

APPENDIX 2

Market Reform and Regulation

Table A2.1 National Telecommunications Laws, as of 2008

	Main law governing telecom sector	*Date*	*Remarks*
Algeria	Loi no. 2000-03 du 05 Joumada El Oula 1421 correspondant au 05 août 2000 fixant les règles générales relatives à la poste et aux télécommunications.	2000	
Angola	Lei no. 8/01 de 11 de Maio	2001	Revokes all legislation contrary to this law, notably Law 4/85 and Decree 19/87
Benin	Ordonnance no. 2002-002 du 31 décembre 2002 portant principes fondamentaux du régime des télécommunications au Bénin	2002	
Botswana	Telecommunications Act	1996	
Burkina Faso	Loi no. 051/98/AN portant reforme du secteur des télécommunications au Burkina Faso	1998	Supersedes earlier telecom regulations but is first law devoted to the sector
Burundi	Décret-Loi no. 1/011 du 4.09.1997	1997	
Cameroon	Loi no. 98/014	1998	Repeals all previous provisions covering telecommunications
Cape Verde	Decreto-Legislativo no. 7/2005 de 24 de Novembro	2005	Revokes Decreto lei no. 5/94 and modifies the concession agreement between the government and Cape Verde Telecom
Chad	Loi no. 009/MPT/98 portant sur les Télécommunications	1998	Supersedes earlier telecom regulations
Central African Republic	Loi no. 96.008 du 13 janvier 1996	1996	
Comoros	Loi no. 97/004/AF du 24 juillet 1997	1997	
Congo, Dem. Rep.	Loi-cadre no. 013/2002 du 16 octobre 2002	2002	Replaces all earlier regulations including the 1940 ordinance covering telecommunications
Congo, Rep.	Loi no. 14-97 du 26 Mai 1997 portant Réglementation du secteur des Télécommunications	1997	Supersedes earlier legislation contrary to the new law
Côte d'Ivoire	Loi no. 95-526 du 7 juillet 1995 Portant Code des Télécommunications	1995	Replaces 1976 law
Djibouti	Loi no. 13/AN/98/4ème portant réforme du Secteur des Postes et Télécommunications	1998	

(*continued next page*)

Table A2.1 *(continued)*

	Main law governing telecom sector	*Date*	*Remarks*
Egypt, Arab Rep.	Telecommunication Regulation Law	2003	
Equatorial Guinea	Ley general de telecomunicaciones	1985	
Eritrea	Communications Proclamation no. 102/98	1998	
Ethiopia	Telecommunications Proclamation no. 49/1996	1996	Repeals several proclamations dating back to 1940s and 1950s
Gabon	005/2001: Loi portant réglementations du secteur des télécommunication	2001	
Ghana	National Communications Authority Act of 2008, Act 769	2008	Repeals the NCA Act of 1996
Kenya	Kenya Communications Act, 1998	1998	Repeals the Kenya Posts and Telecommunications Act
Lesotho	Lesotho Telecommunications Authority Act 2000	2000	Repeals the 1979 Telecommunications Act
Liberia	An Act to Amend the Public Authorities Law creating the Ministry of Posts and Telecommunications, and to Establish an Interim Framework for Telecommunications Regulation	2005	
Madagascar	Loi no. 2005-023	2005	Supersedes all earlier legislation to the contrary including Loi no. 96-034 of 1997
Malawi	Communications Act, 1998	1998	Replaces the 1994 Malawi Posts and Telecommunications Act
Mali	Ordonnance no. 99-043/P-RM du 30 septembre 1999	1999	
Mauritania	Loi no. 99-019 portant sur les telecommunications	1999	
Mauritius	Information and Communication Technologies Act	2001	
Morocco	Dahir no. 1-97-162 du 7/8/97 portant promulgation de la loi no. 24/96 relative à la poste et aux télécommunications	1997	
Mozambique	Lei das Telecomunicações no. 8/2004	2004	Supersedes 1999 Telecommunications Law

(continued next page)

Table A2.1 *(continued)*

	Main law governing telecom sector	*Date*	*Remarks*
Namibia	Namibian Communications Commission Act 1992	1992	Although there have been several amendments, the 1992 Act remains in force
Niger	Ordonnance no. 99-045 du 26 Octobre 1999 Portant réglementation des Télécommunications	1999	Replaces all earlier regulations to the contrary as well as the 1996 Ordinance regulating telecommunications
Nigeria	Nigerian Communications Act 2003	2003	Supersedes previous act of 1992
Rwanda	Law no. 44/2001 of 30/11/2001 Governing Telecommunications	2001	All previous legal provisions contrary to the new law are abrogated including law no. 8/92 relating to Institutional Reform of telecommunications
São Tomé and Príncipe	Lei no. 3/04 Lei que define as regras aplicáveis ao estabelecimento, à gestão e à exploração de redes de telecomunicações nacionais e ao fornecimento de serviços de telecomunicações	2004	Revokes all previous laws contrary to this one
Senegal	Loi no. 2001-15 du 27 décembre 2001 portant code des télécommunications	2001	Supersedes the 1996 Telecommunications Code
Seychelles	Broadcasting and Telecommunication Act	2000	
South Africa	The Electronic Communications Act, no. 36 of 2005	2006	Replaces 1996 Telecommunications Act and large sections of the 1999 Broadcasting Act
Sudan	Telecommunication Act 2001	2001	Repeals the 1974 Telecommunications Act Separate law in the South of Sudan (The Telecommunication Corporation Act, 2004). The relationship between this and the national Telecommunication Act 2001 is currently unclear.

(continued next page)

Table A2.1 *(continued)*

	Main law governing telecom sector	*Date*	*Remarks*
Swaziland	Post and Telecommunications Corporation Act	1983	
Tanzania	Tanzania Communications Act of 1993 and	1993	Largely amends the 1993 Act
	The Tanzania Communications Regulatory Authority Act, 2003	2003	
Togo	Loi no. 98-005 du 11 février 98 sur les télécommunications	1998	Revokes earlier legislation contrary to the present law, notably loi no. 89–14 and ordinance no. 12 of 1974
Tunisia	Loi no. 1-2001 du 15 janvier 2001 portant promulgation du code des télécommunications	2001	
Uganda	The Uganda Communications Act	2000	Although passed in 1997, only entered into force in 2000. Supersedes various telecom legislation dating back to 1970.
Zambia	Telecommunications Act no. 23 of 1994	1994	Replaces the 1987 Posts and Telecommunications Act
Zimbabwe	Postal and Telecommunications Act	2000	

Source: AICD.
Note: Countries for which data are not available are not listed.

Table A2.2 Status of VoIP Services, by Country, 2008

	Status
Algeria	The provision of VoIP requires authorization. The modalities for authorization were issued in a 2005 regulation.[a] Almost a dozen companies have been awarded VoIP licenses.[b]
Angola	Licensed telecommunication operators can provide VoIP.
Benin	Illegal
Botswana	Legal since August 2006; ISPs are allowed to provide VoIP.
Burkina Faso	Tolerated
Cameroon	Illegal
Cape Verde	Legal since 2008
Chad	Illegal
Comoros	Not legal (incumbent has monopoly for voice services)
Congo, Dem. Rep.	VoIP can be provided by licensed facilities telecom providers including payphone operators.

(continued next page)

Table A2.2 *(continued)*

	Status
Côte d'Ivoire	Licensed facilities-based operators are allowed to provide VoIP services.
Djibouti	Djibouti Telecom has a monopoly on voice services.
Egypt, Arab Rep.	"The NTRA has already developed rules and procedures for the licensing of Voice over Internet Protocol (VoIP) services following detailed consultation and discussion over policies, regulations and technical issues. With the liberalization of international gateways, it is expected that many companies will be attracted to the prospects of investing in international VoIP services within the Egyptian market."[c]
Equatorial Guinea	Although illegal, VoIP is tolerated. The incumbent operator has a clause in the Internet contract signed with customers that prohibits the use of VoIP for commercial purposes. However, given there is no specific VoIP legislation and that the incumbent tolerates use, the regulator takes a hands-off approach.
Eritrea	Illegal
Ethiopia	The provision of voice and fax services over the Internet is explicitly prohibited by Proclamation No. 281/2002.
Gabon	Illegal
Gambia, The	Illegal
Ghana	Ghana has yet to legalize VoIP services. The Director General of the National Communications Authority, the regulator, and the Minister of Communications have reportedly publicly stated their readiness to license VoIP operators.[d]
Guinea	Illegal
Guinea-Bissau	Tolerated
Kenya	The Kenyan regulator, the Communications Commission of Kenya (CCK), has issued guidelines legalizing various categories of VoIP, following public consultation.[e] As a result, most operators are carrying VoIP traffic, including telecenters connected to licensed operators.
Lesotho	No apparent restrictions with the expiration of Lesotho Telecom's exclusivity in February 2007. ISPs are legally allowed to own their own international gateways.
Liberia	In the absence of any specific law, VoIP is tolerated.
Malawi	ISPs can lease international data gateways.
Mali	In the absence of any specific law, VoIP is tolerated.
Mauritius	VoIP is legal with two types of licenses: (1) International Long Distance (ILD) license (where calls can be terminated on a PSTN/PLMN telephone) and (2) Internet Telephony Service (where calls cannot be terminated on a PSTN/PLMN telephone in Mauritius).

(continued next page)

Table A2.2 *(continued)*

	Status
Morocco	Only licensed telephone operators can offer service. However, VoIP is authorized for private networks and public call offices that have made a declaration to the regulator.
Mozambique	Legal status is unclear because the regulatory framework does not specifically cover VoIP and private use is tolerated. According to a project implementing a VoIP network in the main university, "IP to IP Voice over IP, as implemented in the UEM intranet, would not be against the law as long as it is not offered as a commercial service."[f]
Namibia	VoIP requires a voice license that only Telecom Namibia has. VoIP operators have been arrested.[g]
Niger	Use of VoIP is allowed with an authorization.
Nigeria	A license is required to provide VoIP services.[h]
Rwanda	Use of VoIP is tolerated.
Senegal	Senegal has not yet defined any specific regulation of VoIP, according to the incumbent, Sonatel. There are many Internet users using Internet telephony over Skype.[i] Individuals can use Skype for their personal communications, but businesses and firms cannot.
South Africa	As of 1 February 2005, any holder of a value added network service or enhanced service license is allowed to carry voice on their networks. VANS are still required to obtain facilities from any licensed telecom operator. ISPs and VANS operators offer VoIP services on a retail basis.
Sudan	An international gateway license Is required to provide VoIP. Restrictions on International gateway licenses were lifted in 2005 and have been issued to facilities-based fixed and mobile operators.
Swazlland	Illegal
Tanzania	The provision of IP telephony is included under the scope of an Application Service License.
Togo	ISPs can lease international data gateways.
Tunisia	VoIP is permitted for business use by companies and firms, for which an authorization is needed from the Tunisian Ministry of Communication Technologies. Tunisie Telecom does not allow VoIP for residential customers. The Ministry is revising the regulatory framework to promote the development of VoIP.
Uganda	Licensed telecom providers are permitted to provide IP telephony on the basis that it is a voice service. ISPs can lease international data gateways.

(continued next page)

Table A2.2 *(continued)*

	Status
Zambia	According to the ICT Policy, "The international gateway for transmission and receiving of telephone calls is another area restricted to Zamtel. This includes Voice over Internet Protocol telephony for commercial purposes."
Zimbabwe	Legal with an Internet Access Provider Class A license

Sources: Adapted from International Telecommunication Union (ITU), Balancing Act, AICD database.

Note: ISP = Internet service provider; NTRA = National Telecommunication Regulatory Authority; PLMN = public land mobile network; PSTN = public switched telephone network; VANS = value-added network service; VoIP = Voice over Internet Protocol; Zamtel = Zambia Telecommunications Company. Countries for which data are not available are not listed.

a. http://www.arpt.dz/Docs/2Textes/Decision/DEC_N04-05.pdf.

b. http://www.arpt.dz/5VOIP.htm.

c. http://www.egyptitutelecom.gov.eg/NTRA.html.

d. Cohen and Southwood 2004.

e. Presentation by John Waweru, Director General and CEO, Communications Commission of Kenya, to the Global Symposium of Regulators 2005, presented at Hammamet in November 2005, as reported in http://www.itu.int/ITU-D/treg/Events/Seminars/2005/GSR05/Documents/chairmansreport.pdf. Also see the consultation at http://www.cck.go.ke/html/child.asp?title=Public+Consultation&contcatid=11&childtitle=Current+Consultations&childcontid=56.

f. http://csd.ssvl.kth.se/~csd2005-team3/docs/MoVoIX-final_report-csd2005.pdf.

g. http://www.telecom.na/index.php?go=news&sel=view&nid=45.

h. http://www.balancingact-africa.com/news/back/french_july04.html#head.

i. "Debate: Is It Necessary to Regulate VoIP?" *Balancing Act Africa* (French edition) no. 21, August 2005, http://www.balancingact-africa.com/news/back/french_august05.html.

Table A2.3 Year of Termination of Exclusivity for Incumbent Telecom Operator, 2002–07

	2002	*2003*	*2004*	*2005*	*2006*	*2007*	*Comment*
Algeria			○	•	•	•	A March 2003 regulatory decision established the opening date for competition in wireless local loop services as of March 2004. A second national operator license was awarded to Consortium Algérien des Télécommunications in 2005.
Angola			•	•	•	•	The 2001 Basic Law on Telecommunications states that the incumbent operator is responsible for ensuring nationwide coverage to fixed telephone services under exclusivity according to contract. However, the law also states that the incumbent operator can compete with other operators to provide public telecommunication services. Four fixed services licenses were awarded in 2004.
Benin				○	○	○	According to the licenses issued to mobile operators, Benin Telecom's exclusivity expired at the end of 2005, allowing the mobile operators to establish their own gateways. However, no additional facilities-based licenses have been issued. The government has reinstated at the end of 2006 a monopoly on international services. Also, all mobile operators have to interconnect with the incumbent to terminate a call outside their own network.
Botswana							In a June 2006 statement, the regulator published deadlines for market liberalization including international gateways as of October 2006 for existing operators and new national licenses as of July 2009.
Burkina Faso					○	○	The 1998 *Telecom Law* granted exclusivity for fixed and international services until the expiration of the concession (in 2006). Mobile operators can operate their own international gateway.

(*continued next page*)

Table A2.3 *(continued)*

	2002	*2003*	*2004*	*2005*	*2006*	*2007*	*Comment*
Cameroon							A decree established a monopoly over long-distance communication. However, after the completion of the planned privatization transaction, the sector policy mentions that the exclusivity will be limited to the resale of national capacity over fiber-optic backbones until 2010.
Cape Verde					○	○	Decreto-Legislativo no. 7/2005 modified the Concession Agreement that had granted CVT a monopoly for the operation of basic services and exclusivity for international communications for 25 years, that is, until January 1, 2021. The exclusivity for leased lines and international services terminated on January 1,2006, and for fixed telephone services on January 1, 2007.
Chad		○	○	○	○	○	According to the 1998 Telecom Law, the incumbent was granted a five-year exclusivity.
Congo, Dem. Rep.	○	○	○	○	○	○	The 2002 Telecom Law opens all markets to competition.
Congo, Rep.							The 1997 Telecom Law states that fixed telephone service is a monopoly.
Côte d'Ivoire			○	○	•	•	CI Telecom's exclusivity over fixed-line infrastructure expired in 2004. A second fixed operator has been licensed.
Djibouti							According to the 1998 Telecom Law, Djibouti Télécom has a "monopoly on telecommunications activities."
Egypt, Arab Rep.					○	○	According to the 2003 Telecom Law, Telecom Egypt had the "exclusive right to establish, operate and exploit international transmission networks between Egypt and any other country through international gateways" up to December 31, 2005.
Ethiopia							ETC is identified as the "sole" telecommunications provider in legal texts.
Gabon						○	The 2001 telecommunications law states that the state could grant exclusivity for up to five years from the publication of the law.

Gambia, The							According to the ICT Bill, "Except where a licence had been issued and exclusivity rights vested in a licensee prior to the enactment of this Act, the Department of State shall not include in a licence or the terms of a licence an exclusivity period or monopoly to the licence."
Ghana	○	○	○	○	○	○	The exclusivity of Westel and Ghana Telecom expired in 2002 and a new licensing framework was introduced in 2003.
Kenya			○	○	○	•	Telkom Kenya's remaining exclusivities came to an end in June 2004. Mobile operators received international licenses in 2006.
Lesotho						○	Exclusivity extended for one year from February 2006.
Madagascar			○	○	○	○	Telma's exclusivity expired in 2004. It was issued a new, nonexclusive license that year.
Malawi					○	○	The 1998 *Communications Sector Policy Statement* granted a five-year exclusivity for fixed and international calls to Malawi Telecom from the date of its incorporation. The statement also noted: "The Government may consider an extension of this exclusivity initially up to one year." MTL was incorporated on May 31, 2000, implying that the exclusivity could have ended in 2005 or 2006 depending on interpretation. However, there apparently is no obligation to issue additional licenses.
Mali		•	•	•	•	•	The 1999 Ordinance covering telecommunications in Mali makes no mention of exclusivity and states that telecommunication networks shall be freely established with a license. Ikatel holds a full service license and launched service as a second national operator in 2003.
Mauritania			○	○	•	•	The historic operator had an exclusive license until June 30, 2004. Chinguitel was awarded a full service license in 2006.

(*continued next page*)

Table A2.3 *(continued)*

	2002	*2003*	*2004*	*2005*	*2006*	*2007*	*Comment*
Mauritius			○	•	•	•	In November 2001, the government announced that it was advancing liberalization to begin from January 1, 2003, with Mauritius Telecom reimbursed for early termination. A fixed telephone license and several international licenses have been issued.
Morocco							A 2004 decree appears to overturn any exclusivities of Maroc Telecom. Two fixed-line licenses have been awarded.
Mozambique							The *Telecommunications Law* grants the incumbent TDM exclusivity for fixed telephone service, which must remain in place until at least December 31, 2007.
Namibia							Although the NCC Act of 1992 theoretically allows the regulator to issue licenses, apart from mobile, no facilities-based voice licenses have been issued. Therefore, Telecom Namibia has a de facto rather than de jure monopoly.
Niger			○	○	○	•	According to the 1999 *Telecommunication Policy Sector Declaration*, Sonitel's exclusivity expires in 2004. A second global fixed-mobile-internet license was awarded in 2007.
Nigeria					○	•	Nigeria has introduced a unified licensing framework, and there are numerous fixed telephone and international operators.
Rwanda							There is no de jure exclusivity.
São Tomé and Príncipe					○	○	According to Lei no. 3/04 of 2004, incumbent CST has international and mobile exclusivity until December 31, 2005.
Senegal			○	○	○	•	Sonatel's exclusivity expired in July 2004. A second global license was awarded in 2007.
South Africa	○	○	○	•	•	•	Telkom's exclusivity expired in May 2002, and a second national operator license was issued in December 2005.
Sudan				○	•	•	Exclusivity over international facilities ended in October 2005. A second national license was awarded to Canar Telecom.

Swaziland							According to the *Post and Telecommunications Corporation Act* of 1983, the Swaziland Posts and Telecommunications Corporation (SPTC) "shall have the exclusive privilege of providing telephone services and of constructing, maintaining and operating telephone apparatus within Swaziland." The act does allow SPTC to award licenses to others to provide telecommunications services.
Tanzania				o	•	•	In February 2005, TCRA adopted a converged licensing framework upon lapse of TTCL's exclusivity.
Togo							The 1998 Telecom Law specifically mentions that Togo Telecom must abide by the new law. The law states that anticompetitive behavior and abuse of market position are forbidden. Authorization to provide service rests with the minister responsible for telecommunications, who can limit authorizations but only in the case of scarce frequency resources.
Tunisia		o	o	o	o	o	The 2001 Telecommunications Code states that no license may be exclusive and furthermore states that existing operators had two years to regularize their situation. This implies that any exclusivity Tunisie Telecom had expired January 2003. A second fixed license was awarded in 2009.
Uganda	⊙	⊙	⊙	⊙	•	•	The duopoly expired in July 2005, and applications for new licenses were accepted from August 2006 with the new license framework operational from January 2007.
Zambia							Zamtel does not appear to have de jure exclusivity for fixed and international services. No additional licenses in these areas have been awarded. There is de facto exclusivity, in particular for international gateways.
Zimbabwe							The Post and Telecommunications Act makes no explicit mention of any exclusivity.

Source: AICD.

Note: o = End of market exclusivity for incumbent operator; • = Introduction of new operators/open market license framework; ⊙ = Regulatory duopoly; CST = Companhia Santomense de Telecomunicacaoes; CVT = Cabo Verde Telecom; ETC = Ethiopian Telecommunications Corporation; TCRA = Tanzania Communications Regulatory Authority; TDM = Telecomunições de Moçambique; TTCL = Tanzania Telecommunications Company Ltd. Countries for which data are not available are not listed.

Table A2.4 Status of Market Liberalization in Sub-Saharan Africa, 2009

	Number of mobile operators	*HHI, 2009, except where otherwise noted*
Comoros	1	10,000
Eritrea	1	10,000
Ethiopia	1	10,000
São Tomé and Principe	1	10,000
Swaziland	1	10,000
Angola	2	5,638
Cameroon	2	5,098
Cape Verde	2	7,214
Chad	2	5,037
Equatorial Guinea	2	10,000
Lesotho	2	6,800
Malawi	2	5,626
Mali	2	6,800
Mozambique	2	5,050
Namibia	2	6,717
Togo	2	5,724 [2008]
Botswana	3	4,432
Burkina Faso	3	4,047
Congo, Rep.	3	4,425
Gabon	3	4,584
Gambia, The	3	n.a.
Guinea-Bissau	3	6,250
Madagascar	3	3,528
Mauritania	3	5,092 [2008]
Mauritius	3	5,160
Rwanda	3	7,828
Senegal	3	5,408
Seychelles	3	n.a.
South Africa	3	4,108
Sudan	3	4,402
Zambia	3	5,478
Zimbabwe	3	4,977 [2008]
Burundi	4	5,276 [2007]
Central African Republic	4	n.a.
Kenya	4	6,630
Liberia	4	n.a.
Niger	4	4,890
Sierra Leone	4	3,522
Benin	5	2,411
Congo, Dem. Rep.	5	3,242
Côte d'Ivoire	5	2,856
Ghana	5	3,332

(continued next page)

Table A2.4 *(continued)*

	Number of mobile operators	*HHI, 2009, except where otherwise noted*
Guinea	5	n.a.
Nigeria	5	3,527
Tanzania	5	3,018
Uganda	5	4,384
Somalia	7	n.a.
Mayotte	n.a.	n.a.

Source: Ampah and others 2009, updated.
Note: HHI = Herfindahl-Hirschman Index; n.a. = not applicable.

Table A2.5 Average Annual Increase in Subscription between Entries of Successive Mobile Operators, through 2008
subscribers per 100 population

	Between first and second operators	*Between second and third operators*	*Between third and fourth operators*	*Between fourth and fifth operators*
Benin	0.03		1.46	8.58
Burkina Faso	0.05		1.43	
Cameroon	0.005		4.51	
Cape Verde	2.50			
Chad		0.74		
Congo, Dem. Rep.	0.00		0.01	2.11
Côte d'Ivoire			3.36	
Ethiopia	0.16			
Ghana		0.04	0.21	4.03
Kenya	0.02	5.25	12.00	
Lesotho		0.52	2.11	
Madagascar			0.01	0.67
Malawi	0.03	0.71		
Mozambique	0.28	3.34		
Namibia	2.65	10.03		
Niger	0.01	0.76	6.54	
Nigeria		0.002	0.89	5.13
Rwanda	0.34	5.32		
Senegal	0.08	3.89		
South Africa	3.55	11.01		
Sudan	0.38	6.56		
Tanzania		0.04		3.15
Uganda	0.01	0.21	1.72	8.61
Zambia	0.01	0.04	2.06	
Average (median)	0.05	0.75	1.72	4.03

Source: AICD.

Table A2.6 Privatizations of Incumbent Operators in Sub-Saharan Africa, 1993–2009

		Initial privatization transaction				
	Operator	*Date*	*% sold*	*Amount ($, million)*	*% private (2008)*	*Notes*
Burkina Faso	ONATEL	Dec. 2006	51	295	51	Private sale to Maroc Telecom
Cape Verde	Cabo Verde Telecom	Dec. 1995	40	20	59	Private sale to Portugal Telecom. Subsequent distribution to employees (5% of total), national private investors (14%), and government social security system (38%).
Central African Republic	SOCATEL					France Cable and Radio owned 40% of shares at one point. Current status not available.
Côte d'Ivoire	Côte d'Ivoire Telecom	Jan. 1997	51	210	51	Private sale to France Telecom
Equatorial Guinea	Getesa	1987	40		40	Private sale to France Telecom
Gabon	Gabon Telecom	Feb. 2007	51	79	51	Private sale to Maroc Telecom
Gambia	GAMTEL	Jan. 2007	50	35	50	Private sale to Spectrum Investment Holding (Lebanon)
Ghana	Ghana Telecom	Dec. 1996	30	38	70	Original private sale to G-Com consortium headed by Telekom Malaysia. In 2002, the government of Ghana abrogated the management contract with G-Com and bought back shares. Subsequent private sale of 70% to Vodafone (United Kingdom) in August 2008 for $900 million.
Guinea	SOTELGUI	Dec. 1995	60	45	0	Renationalized in 2008 following private sale to Telekom Malaysia
Guinea-Bissau	Guinée Telecom	1989	51	3	0	Renationalized following private sale to Marconi (later assumed by Portugal Telecom)
Kenya	Telkom Kenya	Dec. 2007	51	390	51	Sale to consortium led by France Telecom (78.5%) with Alcazar Capital Limited (21.5%)

Lesotho	Telecom Lesotho	Nov. 2000	70	—	70	Private sale to Mountain Communications (Econet, Zimbabwe), Mauritius Telecom, and Eskom (South Africa). Sale price not disclosed.
Madagascar	TELMA	Aug. 2003	34	13	68	Private sale to Distacom (Hong Kong, China), which also purchased France Telecom's ownership
Malawi	Malawi Telecom	Feb. 2006	80	30	80	Private sale to THL consisting of PCL (50.1%), Old Mutual (16.08%), NICO (5%), Detecon, Germany (2.603%), and Press Trust (6.217%). Percentages refer to MTL ownership.
Mali	SOTELMA	July 2009	51	384	51	Private sale to Maroc Telecom
Mauritania	MAURITEL	Apr. 2001	54	48	54	Private sale to Maroc Telecom, which subsequently engaged in a series of sales with local investors. Its ownership stood at 51% in 2008.
Mauritius	Mauritius Telecom	Nov. 2000	40	261	40	Private sale to France Telecom
Niger	SONITEL	Nov. 2001	51	16	51	Private sale to a Chinese and Libyan consortium. Government has announced intention to renationalize.
Nigeria	NITEL	July 2006	51	500		Private sale to TransCorp (Nigeria). Government has rescinded the sale and was in the process of reprivatizing in 2010.
Rwanda	Rwandatel	June 2005	99	20	80	Initial private sale to Terracom (United States), which government later repurchased. LAPGreen Networks later acquired an 80% interest for $100 million.
São Tomé and Príncipe	CST	1989	51	1	51	Private sale to Portugal Telecom
Senegal	SONATEL	Jul. 1997	33	90	73	Initial private sale to France Telecom. Subsequent additional sale to France Telecom and listing on regional stock exchanges.

(*continued next page*)

Table A2.7 *(continued)*

	Etisalat (United Arab Emirates)	*France Telecom*	*Maroc Telecom*	*Millicom (Luxembourg)*	*MTC (Zain, Kuwait)*	*MTN (South Africa)*	*Orascom (Egypt, Arab Rep.)*	*Vodafone (United Kingdom)*	*Portugal Telecom*	*Other*	*Total operators*	*Note*
	Percentage ownership											
Côte d'Ivoire	—	85				65				—	4	Other(s) = Comium (Lebanon) and Warid (United Arab Emirates)
Equatorial Guinea		40								—	2	Other = Hits Telecom (Saudi Arabia)
Eritrea											0	
Ethiopia											0	
Gabon	—		51		90						3	
Gambia, The										—	1	Other = Comium (Lebanon)
Ghana				100	75	98		70		—	5	Other = Glo (Nigeria)
Guinea		38				75				—	4	Other = Teylium (Côte d'Ivoire), Cellcom (United States)
Guinea-Bissau		42				100					2	
Kenya		51			80			40		—	4	Other = Essar (India)
Lesotho								88		—	2	
Liberia						60				—	3	Other = Cellcom (United States), Comium (Lebanon)

Madagascar		72			100					68	3	Other = Distacom (Hong Kong, China)
Malawi					100						1	
Mali		30									1	
Mauritania			52							—	2	Other = Tunisia Telecom
Mauritius		40		50							2	
Mozambique								98			1	
Namibia							100		34		2	
Niger	57	80			90					—	4	Other = ZTE (China) and Libyan Arab Portfolio (LAP)
Nigeria	40				66	76					3	
Rwanda						59				—	2	Other = LAP
São Tomé and Príncipe									51		1	
Senegal		42		100							2	
Seychelles										—	2	Other(s) = Cable & Wireless (United Kingdom); Airtel (India)
Sierra Leone					100					—	2	Other = Comium (Lebanon)

(*continued next page*)

Table A2.7 *(continued)*

	Etisalat (United Arab Emirates)	*France Telecom*	*Maroc Telecom*	*Millicom (Luxembourg)*	*MTC (Zain, Kuwait)*	*MTN (South Africa)*	*Orascom (Egypt, Arab Rep.)*	*Vodafone (United Kingdom)*	*Portugal Telecom*	*Other*	*Total operators*	*Note*
	Percentage ownership											
Somalia												
South Africa						100		65		75	3	Other = Oger (Saudi Arabia)
Sudan					100	85					2	
Swaziland						30					1	
Tanzania	—			100	60			65			4	
Togo	—										1	
Uganda		53			100	95				69	5	Other = LAP. In addition, Warid (United Arab Emirates) has stakes in Warid Telecom Uganda.
Zambia					79	100					2	
Zimbabwe							60				1	
Total	9	15	4	6	15	16	7	7	5	12	96	

Source: Ampah and others 2009, updated.

Note: Percentages refer to ownership stakes of strategic investors. The vertical total column refers to the total number of mobile operators in the country with foreign investors, whereas the horizontal total column at the bottom refers to the number of countries each strategic investor has investments in. — = not available. MTC = Mobile Telecommunications Co.; ZTE = Zhong Xing Telecommunication Equipment Company Ltd.

Table A2.8 National Regulatory Authorities, September 2009

	Authority	*Year created*	*Web site*	Responsible for post?	Multisectoral?	Number of staff
Algeria	Autorité de Régulation de la Poste et des Télécommunications (ARPT)	2000	http://www.arpt.dz	Yes	No	115
Angola	Instituto Angolano das Comunicações	1999	http://www.inacom.og.ao	No	No	n.a.
Benin	n.a.	n.a.	n.a.	n.a.	n.a.	n.a.
Botswana	Botswana Telecommunications Authority	1996	http://www.bta.org.bw	No	No	71
Burkina Faso	Autorité Nationale de Régulation des Télécommunications (ARTEL)	1998	http://www.artel.bf	No	No	27
Burundi	Agence de Régulation et de Contrôle des Télécommunications	1997	—	No	No	—
Cameroon	Agence de Régulation des Télécommunications	1998	http://www.art.cm	No	No	—
Central African Republic	Agence Chargée de la Régulation des Télécommunications (ART)	1996	http://www.art-rca.org	No	No	29
Cape Verde	Agência Nacional de Comunicações (ANAC)	2004	http://www.icti.cv	Yes	No	10
Chad	Office Tchadien de Régulation des Télécoms (OTRT)	1998	http://www.otrt.td	No	No	50
Comoros	n.a.	n.a.	n.a.	n.a.	n.a.	n.a.
Congo, Dem. Rep.	Autorité de Régulation de la Poste et des Télécommunications du Congo (ARPTC)	2002	http://www.arptc.cd	Yes	No	32
Congo, Rep.	Direction de l'Administration Centrale des Postes et Télécommunications (DGACPT)	2003	http://www.dgacpt.com	Yes	No	—

(*continued next page*)

Table A2.8 *(continued)*

	Authority	*Year created*	*Web site*	*Responsible for post?*	*Multisectoral?*	*Number of staff*
Côte d'Ivoire	Agence des télécommunications de Côte d'Ivoire (ATCI)	1995	http://www.atci.ci	No	No	130
Djibouti	n.a.	n.a.	n.a.	n.a.	n.a.	n.a.
Egypt, Arab Rep.	National Telecommunication Regulatory Authority (NTRA)	2003	http://www.tra.gov.eg	No	No	345
Equatorial Guinea	Oficina Reguladora de Telecomunicaciones (ORTEL)	—	—	No	No	11
Eritrea	Communications Department	1998	—	—	No	14
Ethiopia	Ethiopian Telecommunications Agency (ETA)	1996	http://www.eta.gov.et	No	No	43
Gabon	Agence de Régulation des Télécommunications (ARTEL)	2001	http://www.artel.ga	No	No	137
Gambia	Gambia Public Utilities Regulatory Authority	2004	http://www.pura.gm	[a]	Yes	31
Ghana	National Communications Authority (NCA)	1997	http://www.nca.org.gh	No	No	48
Guinea	Autorité de Régulation des Postes et Télécommunications (A.R.P.T.)	2008	—	Yes	No	—
Guinea-Bissau	Institut des Communications de la Guinée-Bissau (ICGB)	1999	—	Yes	No	—
Kenya	Communications Commission of Kenya (CCK)	1999	http://www.cck.go.ke	Yes	No	130
Lesotho	Lesotho Telecommunications Authority	2000	http://www.lta.org.ls	No	No[b]	25
Liberia	Liberia Telecommunications Authority (LTA)	2005	—	No	No	11
Libya	—	—	—	—	—	—
Madagascar	Office Malagasy d'Etudes et de Régulation des Télécommunications (OMERT)	1997	http://www.omert.mg	No	No	60

Malawi	Malawi Communications Regulatory Authority (MACRA)	1998	http://www.macra.org.mw	Yes	No	26
Mali	Comité de Régulation des Télécommunications (CRT)	1999	http://www.crt-mali.org	No	No	17
Mauritania	Autorité de Régulation (ARE)	1999	http://www.are.mr	Yes	Yes	45
Mauritius	Information and Communication Technologies Authority (ICTA)	2002	http://www.icta.mu	No	No	50
Morocco	Agence Nationale de Réglementation des Télécommunications (ANRT)	1997	http://www.anrt.ma	No	No	182
Mozambique	Instituto Nacional das Comunicações de Moçambique (INCM)	2002	http://www.incm.gov.mz	Yes	No	84
Namibia	Namibian Communications Commission (NCC)	1992	http://www.ncc.org.na	No	No	7
Niger	Autorité de Régulation Multisectorielle (ARM)	2004	http://www.arm-niger.org	No	Yes	25
Nigeria	Nigerian Communications Commission (NCC)	1992	http://www.ncc.gov.ng	No	No	222
Rwanda	Rwanda Utilities Regulatory Agency (RURA)	2001	http://www.rura.gov.rw	No	Yes	34
São Tomé and Príncipe	Autorité Générale de Régulation	2005	—	Yes	Yes	20
Senegal	Agence de Régulation des Télécommunications et des Postes (ARTP)	2001	http://www.artp-senegal.org	Yes	No	55
Seychelles	n.a.	n.a.	n.a.	n.a.	n.a.	n.a.
Sierra Leone	National Telecommunications Commission	2006	—	No	No	35
Somalia	n.a.	n.a.	n.a.	n.a.	n.a.	n.a.
South Africa	Independent Communications Authority of South Africa (ICASA)	2000	http://www.icasa.org.za	[a]	No[b]	316
Sudan	National Telecommunication Corporation (NTC)	1994[c]	http://www.ntc.org.sd	No	No	100
Swaziland	n.a.	n.a.	n.a.	n.a.	n.a.	n.a.

(continued next page)

Table A2.9 *(continued)*

	Are interconnection agreements made public?	*Are interconnection prices made public?*	*Are operators required to publish a reference interconnection offer (RIO)?*
Sierra Leone[a]	No	No	—
Somalia	—	—	—
South Africa[a]	Yes	Yes	No
Sudan	No	No	Yes
Swaziland[a]	No	No	—
Tanzania[a]	No	Yes	No
Togo[a]	No	Yes	Yes
Uganda[a]	Yes	Yes	No
Zambia[a]	No	No	No
Zimbabwe[a]	No	No	Yes

Source: Ampah and others 2009.
Note: — = not available.
a. Pre-2007 data.

Table A2.10 Status of Universal Service Funds, by Country, 2009

	Universal service fund in operation?	*Fund*	*Date established*	*Contribution by licensed operators*	*Note*
Algeria	Yes	—		3% of revenue	
Angola	Yes	—			The 2001 sector policy calls for the "Fundo de Apoio ao Desenvolvimento das Comunicações (FADCOM)" to finance universal service.
Benin	No				
Botswana	No				
Burkina Faso	Yes	Fonds d'accès au service universel des télécommunications	2000	2.5% of revenue	
Burundi	No				
Cameroon	Yes	—			
Cape Verde	No				
Central African Republic	No				
Chad	Yes	—			
Comoros	No				
Congo, Dem. Rep.	No				
Congo, Rep.	No				
Côte d'Ivoire	Yes	Fonds National des telecommunications (FNT)	1998		

(continued next page)

Table A2.10 *(continued)*

	Universal service fund in operation?	*Fund*	*Date established*	*Contribution by licensed operators*	*Note*
Djibouti	No				
Egypt, Arab Rep.	Yes	Universal Service Fund	2005		See http://www.tra.gov.eg/english/DPages_DPagesDetails.asp?ID=226&Menu=1.
Equatorial Guinea	No				
Eritrea	No				
Ethiopia	No				
Gabon	Yes	Fonds Spécial du Service Universel (FSSU)	2007	2% of revenue	All licensed operators except incumbent contribute.
Gambia	No				
Ghana	Yes	Ghana Investment Fund for Telecommunications (GIFTEL)	2005	1% of annual net revenue	See http://giftel.gov.gh.
Guinea	Yes	—			
Guinea-Bissau	No				
Kenya	No				
Lesotho	Yes	Universal Access Fund	2009	1% of net income	
Liberia	No				
Libya					
Madagascar	Yes	Fonds de développement des télécommunications et TIC	1999	2% of revenue after taxes	Ariary 1 billion spent in 2006 on reimbursing incumbent for telephone service in 40 localities and another operator for VSAT network in 11 localities

Malawi	No				
Mali	No				
Mauritania	Yes	Fonds d'Accès Universel (FAU)	2002	3% of revenue	
Mauritius	Yes	—	2008		
Morocco	Yes	—	2005	2% of revenue	Used to finance GSM coverage expansion, provision of Internet in schools and telecenters in rural areas
Mozambique	No				
Namibia	No				
Niger	Yes	Fonds d'Accès Universel		4% of gross revenue	
Nigeria	Yes	Universal Service Provision Fund	2007	2.5% of gross revenue (less interconnect payments)	See http://www.uspf.gov.ng.
Rwanda	Yes	Universal Access Fund (UAF)	2004	2% of net revenues	RWF 838 million spent in 2008 on providing internet connections to government institutions, police stations, military camps, schools, universities, hospitals, etc.
São Tomé and Príncipe	No				
Senegal	Yes	Fonds de développement du service universel des télécommunications (FDSUT)	2001	3% maximum of revenue excluding tax and interconnection charges	Pilot project in Matam region

(*continued next page*)

Table A2.10 *(continued)*

	Universal service fund in operation?	*Fund*	*Date established*	*Contribution by licensed operators*	*Note*
Seychelles	No				
Sierra Leone	No				
Somalia	No				
South Africa	Yes	Universal Service Fund	1996	0.2% of net revenues	Fund has been disbursing for several years. See http://www.usa.org.za/.
Sudan	Yes	ICT Fund	2003	Levied on tariffs of services	
Swaziland	No				
Tanzania	No				
Togo	Yes	Fonds du Service Universel	2006	2% of revenue	Operators directly reimbursed for investing in unserved areas
Tunisia	Yes	Fonds de développement des communications, des technologies de l'information et de la télécommunication		5% of revenue	
Uganda	Yes	Rural Communication Development Fund	2003	1% of gross revenue	Fund has been disbursing for several years. See http://www.ucc.co.ug/rcdf/default.php.
Zambia	No				
Zimbabwe	Yes	Universal Service Fund	2001	2% gross revenue	

Source: Adapted from ITU and regulator information.
Note: — = information not available.

References

Ampah, Mavis, Daniel Camos, Cecilia Briceño-Garmendia, Michael Minges, Maria Shkaratan, and Mark Williams. 2009. "Information and Communications Technology in Sub-Saharan Africa: A Sector Review." AICD Background Paper 10, World Bank, Washington, DC.

Cohen, T., and R. Southwood. 2004. "An Overview of VOIP Regulation in Africa: Policy Responses and Proposals." *Balancing Act Africa* (London), June.

APPENDIX 3

Financing Telecommunications in Africa

Table A3.1 Investment in Telecommunications Infrastructure, 1998–2008

	Total investment (1998–2008)	
	$, billion	*% of GDP*
Angola	1,054	0.4
Benin	201	0.5
Botswana	298	0.3
Burkina Faso	221	0.4
Burundi	53	0.6
Cameroon	769	0.5
Cape Verde	5	0.0
Central African Republic	3	0.0
Chad	189	0.4
Congo, Dem. Rep.	1,241	1.6
Congo, Rep.	290	0.5
Côte d'Ivoire	751	0.0
Eritrea	40	0.4
Gabon	144	0.2
Gambia, The	7	0.1
Ghana	1,137	1.1
Guinea	155	0.4
Guinea-Bissau	59	2.0
Kenya	2,941	1.5
Lesotho	97	0.8
Liberia	120	2.0
Madagascar	222	0.4
Malawi	146	0.5
Mali	193	0.4
Mauritania	106	0.6
Mauritius	144	0.2
Mozambique	231	0.4
Namibia	53	0.1
Niger	151	0.5
Nigeria	12,759	1.3
Rwanda	128	0.5
Senegal	1,584	1.9
Seychelles	59	0.0
Sierra Leone	144	1.2
Somalia	13	0.0
South Africa	18,130	0.9
Sudan	1,828	0.7
Swaziland	70	0.3
Tanzania	1,405	1.1
Togo	5	0.0
Uganda	1,620	1.8

(continued next page)

Table A3.1 *(continued)*

	Total investment (1998–2008)	
	$, billion	*% of GDP*
Zambia	778	1.1
Zimbabwe	261	0.4
Resource-rich	17,810	0.9
Low-income, nonfragile	10,284	0.9
Low-income, fragile	2,852	1.1
Middle-income	18,855	0.8
Sub-Saharan Africa	49,801	0.9

Source: PPI Database.
Note: Data for the Comoros, Equatorial Guinea, Ethiopia, and São Tomé and Príncipe were not available. GDP = gross domestic product.

Table A3.2 *(continued)*

	Year	*Status*	*Project*	*Chinese financier*	*Contractor*	*Added capacity (thousands of connections)*	*Project cost ($, million)*	*Chinese commitments ($, million)*
Ghana	2005	Completed	Build a CDMA 2000 1X network for Kasapa Telecom	Unknown	ZTE	500	—	—
Ghana	2006	Construction	National Fibre Backbone Project	China Eximbank	Huawei	—	70	31
Ghana	2007	Construction	Communication system for security agencies project	China Eximbank	ZTE	—	—	Unconfirmed
Lesotho	2007	Agreement	Rehabilitate the Telecom Agricultural Network	China Eximbank	ZTE	—	—	30
Lesotho	2007	Construction	Grant to establish television systems in several cities	Ministry of Commerce, China	Unknown	—	—	3
Mali	2005	Agreement	Rehabilitate CDMA2000 1X WLL network in Bamako	ZTE	ZTE	—	2	1
Mauritius	2006	Construction	Milcom purchase by China Mobile	Unknown	China Mobile	250	—	—
Niger	2001	Completed	Equip Niger Telecommunications Company (SONITEL) with GSM mobile system covering the city of Niamey	Unknown	ZTE	—	8	Unconfirmed
Niger	2001	Completed	Tender for 51% ownership of SONITEL, Niger's state telecom company, and its mobile arm, SahelCom	ZTE	ZTE	—	—	24
Nigeria	2002	Construction	National Rural Telephony Project (NRPT), Phase 1	China Eximbank	Huawei; ZTE; ASB	150	200	200

Nigeria	2006	Completed	Nigeria First Communication Satellite NIGCOMSAT	China Eximbank	China Great Wall Industry Corp.	—	—	200
Senegal	2007	Construction	Build the e-government network	China Eximbank	Huawei; China National Machinery & Equipment Corporation (CMEC)	—	51	51
Sierra Leone	2005	Completed	Provision of CDMA fixed-wireless network to government-owned Sierratel	China Eximbank	Huawei	100	17	17
Sierra Leone	2006	Construction	Upgrade the rural telecom network	China Eximbank	Huawei	—	—	18
Sudan	2005	Agreement	Purchase of equipment from ZTE by Sudan Telecom	China Eximbank	ZTE	—	—	200
Togo	2005	Completed	Expansion and upgrade the GSM network of Togo Cellulaire	China Eximbank	ASB	100	17	Unconfirmed
Zambia	2006	Construction	Deploy fiber-optic lines over the Zambia Electricity Supply Corporation Limited (ZESCO) power transmission network	Unknown	ZTE	—	11	—
Zimbabwe	2004	Construction	Two contracts for telecom equipment supply with Zimbabwe's state-owned fixed-line operator TelOne and mobile operator NetOne	China Eximbank	Huawei	—	332	Unconfirmed
Telecom total								2,774

Sources: Foster and others 2008; World Bank 2007.
Note: — = data not available; CDMA = Code Division Multiple Access; GPRS = general packet radio service;

References

Foster, Vivien, William Butterfield, Chuan Chen, and Nataliya Pushak. 2008. "Building Bridges: China's Growing Role as Financier for Africa's Infrastructure." Public-Private Infrastructure Advisory Facility, World Bank, Washington, DC.

PPI (Private Participation in Infrastructure) Database. World Bank and Private Participation in Infrastructure Advisory Facility, Washington, DC. http://www.ppiaf.org/ppiaf/page/private-participation-infrastructure-database.

World Bank. 2007. PPIAF Chinese Projects Database. World Bank, Washington, DC.

APPENDIX 4

Future Investment Needs

Table A4.2 *(continued)*

	Efficient market gap ($ million per year)	*Coverage gap ($ million per year)*	*Investment required to reach universal coverage ($ million per year)*	*Annual investment required to reach universal coverage (% of GDP)*
Madagascar	32.4	53.9	86.3	1.35
Malawi	0.6	5.7	6.3	0.24
Mali	41.4	47.9	89.2	1.29
Mauritania	6.2	20.6	26.8	0.87
Mauritius	0.1	0.0	0.1	0.00
Mozambique	27.7	47.8	75.4	0.85
Namibia	1.2	5.5	6.7	0.09
Niger	12.2	24.7	36.9	0.89
Nigeria	71.7	5.9	77.7	0.06
Rwanda	2.7	0.5	3.2	0.11
São Tomé and Príncipe	0.1	0.5	0.7	0.48
Senegal	8.3	7.1	15.4	0.15
Seychelles	0.0	0.0	0.0	0.00
Sierra Leone	3.5	4.4	7.9	0.47
Somalia	—	—	—	—
South Africa	0.0	3.4	3.4	0.00
Sudan	94.6	54.0	148.5	0.34
Swaziland	1.4	1.3	2.7	0.09
Tanzania	31.2	38.4	69.6	0.47
Togo	5.9	1.6	7.5	0.29
Uganda	1.6	6.9	8.5	0.08
Zambia	31.7	56.0	87.7	0.69
Zimbabwe	11.4	17.4	28.9	0.50
Sub-Saharan Africa	762.2	964.6	1,726.7	0.21
North Africa	105.9	238.2	344.0	0.09
Africa	869.1	1,204.8	2,073.9	0.17

Source: Mayer and others 2009.
Note: — = not available; GDP = gross domestic product.

Table A4.3 Key Cost Assumptions Used in Broadband Infrastructure Modeling

Parameter	*Value or definition*	*Source*
Type of coverage	Outdoor	Mayer and others 2009
Capital expenditure (CAPEX)	$181,000 per four-sector WiMAX 802.16d cell site (single radio per sector, no antenna diversity) $126,000 per single-sector WiMAX 802.16d cell site (single radio per sector, no antenna diversity) $216,000 per CDMA 450 1 × EVDO cell site	Mayer and others 2009

(continued next page)

Table A4.3 *(continued)*

Parameter	*Value or definition*	*Source*
Operational expenditure—network related (network OPEX)	$50,000 per cell site per year	Mayer and others 2009
Nonnetwork OPEX	$50 per subscriber per year	Mayer and others 2009
International bandwidth costs ($)	$2,000 per month for a 2 Mbps international connection (equivalent to $209 per subscriber per year)	Mayer and others 2009
Size of cell sites	Rural = 5,024 (km^2) (radius of 40 km) Urban = 78.5 km^2 (radius of 5 km)	Ho 2005; Seybold 2006
Terrain factor	An integer factor ranging from 1 to 4 that is used to adjust the number of base stations per cell site based on terrain. The factor is calculated based on the percentage of raster cells with unobstructed line of sight to a centrally located high point in the cell site representing a hypothetical antenna position.	Mayer and others 2009. Line-of-sight analysis was conducted using SRTM digital elevation data at 90 meter resolution. ESRI ArcGIS 9.2 software was used.

Sources: As indicated in table.
Note: CDMA = Code Division Multiple Access; EVDO = evolution–data optimized; km = kilometer; km^2 = square kilometer; Mbps = megabits per second; SRTM = shuttle radar topography mission; WiMAX = Worldwide Interoperability for Microwave Access.

Table A4.4 Broadband Infrastructure Modeling Scenarios

Assumption	*Scenario 1: Limited broadband access*	*Scenario 2: Mass-market broadband access*
Subscriber penetration rate	One broadband connection per 100 urban inhabitants plus 1 broadband connection per 400 rural inhabitants	20 broadband connections per 100 urban inhabitants (20% urban penetration) plus 10 broadband connections per 100 rural inhabitants (10% rural penetration)
Revenue generated from broadband	1% of GDP per capita, weighted for urban and rural income distribution. Where data are available, operator revenue is reduced by applicable value added tax and excise taxes (Ampah and others 2009).	Average revenue per subscriber $10 per month

Source: Mayer and others 2009.

Table A4.5 Average Annual Investment Requirement in Shared-Access Scenario

	Efficient market (% of population)	*Coverage gap (% of population)*	*Efficient market ($ million per year)*	*Coverage gap ($ million per year)*	*Investment required to reach universal coverage ($ million per year)*	*Annual investment required to reach universal coverage (% of GDP)*
Angola	94	6	53.9	7.9	61.8	0.12
Benin	95	5	15.6	3.3	18.9	0.34
Botswana	85	15	9.2	8.2	17.4	0.14
Burkina Faso	89	11	19.1	10.5	29.6	0.41
Burundi	54	46	4.0	5.8	9.8	1.03
Cameroon	87	13	39.3	17.9	57.2	0.26
Cape Verde	93	7	2.2	0.9	3.1	0.23
Central African Republic	36	64	5.7	27.3	33.0	1.98
Chad	76	24	16.4	14.0	30.4	0.39
Comoros	94	6	1.8	0.3	2.2	0.46
Congo, Dem. Rep.	0	100	0.0	199.1	199.1	1.98
Congo, Rep.	79	21	10.3	18.8	29.1	0.40
Côte d'Ivoire	95	5	44.1	6.4	50.4	0.24
Equatorial Guinea	100	0	3.1	0.0	3.1	0.03
Eritrea	38	62	3.4	8.9	12.3	1.04
Ethiopia	70	30	60.5	66.8	127.2	0.81
Gabon	88	12	6.9	11.9	18.8	0.18
Gambia, The	37	63	1.8	3.1	4.9	0.87
Ghana	96	4	44.7	5.6	50.3	0.33
Guinea	54	46	12.9	15.1	28.0	0.71
Kenya	94	6	4.2	2.3	6.5	0.37

Lesotho	77	23	0.0	12.9	12.9	2.00
Liberia	0	100	15.7	42.9	58.6	0.90
Madagascar	31	69	10.7	10.8	21.5	0.82
Malawi	58	42	22.8	33.7	56.5	0.81
Mali	73	27	53.9	7.9	61.8	0.12
Mauritania	77	23	8.0	9.7	17.6	0.56
Mauritius	100	0	4.6	0.0	4.6	0.06
Mozambique	63	37	29.7	32.9	62.5	0.70
Namibia	78	22	5.6	12.3	17.9	0.24
Niger	61	39	15.1	17.2	32.3	0.77
Nigeria	100	0	240.2	3.4	243.6	0.18
Rwanda	98	2	8.7	1.0	9.8	0.33
São Tomé and Príncipe	94	6	0.4	0.2	0.6	0.42
Senegal	94	6	24.0	4.7	28.7	0.27
Seychelles	83	17	0.5	0.6	1.1	0.12
Sierra Leone	62	38	7.6	6.4	14.0	0.82
South Africa	99	1	158.1	6.8	164.9	0.05
Sudan	94	6	81.3	28.2	109.5	0.25
Swaziland	100	0	3.4	0.0	3.4	0.11
Tanzania	72	28	60.2	30.9	91.0	0.60
Togo	97	3	11.2	0.8	12.0	0.46
Uganda	93	7	34.2	6.1	40.3	0.37
Zambia	72	28	26.7	22.6	49.3	0.38
Zimbabwe	44	56	19.8	22.9	42.7	0.72
Total	75	25	1,197.6	755.5	1,953.0	0.26

Source: Mayer and others 2009.
Note: GDP = gross domestic product.

References

Ampah, Mavis, Daniel Camos, Cecilia Briceño–Garmendia, Michael Minges, Maria Shkaratan, and Mark Williams. 2009. "Information and Communications Technology in Sub-Saharan Africa: A Sector Review." AICD Background Paper 10, World Bank, Washington, DC.

Ho, K. 2005. "Bridging the Digital Divide: Benefits of Using Lower Frequency Bands for Broadband Wireless Access in Remote Areas." Presentation at CDMA Development Group's "CDMA2000 as a Broadband Access Technology" workshop. http://cdg.org/news/events/CDMASeminar/051115_EVDOWorks.

Mayer, Rebecca, Ken Figueredo, Mike Jensen, Tim Kelly, Richard Green, and Alvaro Federico Barra. 2009. "Connecting the Continent: Costing the Needs for Spending on ICT Infrastructure in Africa." Background Paper 3, Africa Infrastructure Country Diagnostic, World Bank, Washington, DC.

Seybold, Andy. 2006. "Pakistan: A Hotbed of Wireless Activity." *3G Today Newsletter* 3, No. 6 (November). http://www.3gtoday.com/wps/portal/newsletterdetail?newsletterId=1340.

APPENDIX 5

Rates of Penetration of ICT Services in Sub-Saharan Africa

Table A5.1 Fixed Telephone Lines and International Voice Traffic in Sub-Saharan Africa, by Economy and Economy Group, 1998 and 2008

	Total fixed telephone lines		*Fixed telephone lines per 100 inhabitants*		*International voice traffic (minutes per person per month)*	
	1998	*2008 (or latest available)*	*1998*	*2008 (or latest available)*	*1998*	*2008 (or latest available)*
Angola	65,100	114,296	0.5	0.6	3.7	..
Benin	38,354	159,000	0.6	1.8	4.4	11.9
Botswana	102,016	142,282	6.1	7.4	41.6	114.5
Burkina Faso	41,218	144,000	0.4	0.9	2.0	10.9
Burundi	17,849	30,411	0.3	0.4	1.0	..
Cameroon	93,920	198,321	0.6	1.0	4.5	3.6
Cape Verde	39,985	71,860	9.5	14.4	50.8	101.2
Central African Republic	9,563	12,000	0.3	0.3	2.1	..
Chad	8,631	13,000	0.1	0.1	1.0	..
Comoros	6,226	23,300	1.2	3.6	12.5	..
Congo, Dem. Rep.	..	..	..	..	..	..
Congo, Rep.	22,000	22,200	0.8	0.6	..	..
Côte d'Ivoire	170,001	356,502	1.0	1.7	6.3	..
Equatorial Guinea	5,580	10,000	1.1	1.5	..	..
Eritrea	24,308	40,415	0.7	0.8	4.5	17.1
Ethiopia	164,140	897,287	0.3	1.1	0.9	2.1
Gabon	38,698	26,500	3.3	1.8	34.0	69.8
Gambia, The	25,609	48,868	2.1	2.9	..	..
Ghana	133,426	143,900	0.7	0.6	7.0	6.0
Guinea	15,213	21,000	0.2	0.2	3.6	..
Guinea-Bissau	8,079	4,647	0.6	0.3	4.7	..
Kenya	288,251	243,741	1.0	0.6	3.5	3.4
Lesotho	21,000	65,200	1.2	3.2	..	17.9
Liberia	6,500	2,000	0.3	0.1	..	..
Madagascar	47,193	164,851	0.3	0.9	1.8	1.1
Malawi	37,371	175,000	0.3	1.2	1.9	..
Mali	27,063	81,076	0.3	0.6	..	2.4
Mauritania	15,030	76,354	0.6	2.4	6.4	3.7
Mauritius	245,367	364,536	21.1	28.7	57.0	99.7
Mayotte	12,200	10,000	..	5.2	..	..
Mozambique	75,354	78,324	0.4	0.3	35.2	12.5
Namibia	105,877	140,000	6.1	6.6	61.5	..
Niger	18,114	64,738	0.2	0.4	..	..
Nigeria	438,619	1,307,625	0.4	0.9	..	0.7

(continued next page)

Table A5.1 *(continued)*

	Total fixed telephone lines		*Fixed telephone lines per 100 inhabitants*		*International voice traffic (minutes per person per month)*	
	1998	*2008 (or latest available)*	*1998*	*2008 (or latest available)*	*1998*	*2008 (or latest available)*
Rwanda	10,825	16,770	0.2	0.2	..	10.9
São Tomé and Príncipe	4,295	7,700	3.2	4.8	15.7	17.1
Senegal	139,549	237,752	1.5	1.9	13.7	27.5
Seychelles	18,750	22,322	23.8	25.7	129.6	..
Sierra Leone	17,407	31,500	0.4	0.6	..	..
Somalia	23,000	100,000	0.3	1.1	..	..
South Africa	5,075,417	4,425,000	12.1	9.1	22.3	..
Sudan	162,225	366,200	0.5	0.9	2.7	6.0
Swaziland	28,999	44,000	2.8	3.8	49.4	..
Tanzania	121,769	123,809	0.4	0.3	1.0	0.5
Togo	31,415	140,919	0.6	2.2	5.6	5.6
Uganda	56,919	168,481	0.2	0.5	1.1	7.0
Zambia	77,700	90,600	0.8	0.7	3.4	..
Zimbabwe	236,530	348,000	1.9	2.8	8.7	22.3
Economy group						
Low-income, fragile	34,545	51,503	0.37	0.54	2.0	2.9
Low-income, nonfragile	114,702	309,711	0.44	0.76	4.1	3.9
Resource-rich	285,646	813,541	0.44	0.85	1.4	1.9
Middle-income	1,064,090	1,368,695	2.60	2.43	6.7	4.0
CEMAC	44,829	86,404	0.49	0.61	3.9	3.4
EAC	142,790	156,470	0.50	0.45	1.7	3.7
ECOWAS	270,047	757,886	0.52	0.98	2.2	3.4
SADC	1,074,122	899,306	3.00	2.39	10.1	3.7
Sub-Saharan Africa	8,367,075	11,366,287	1.40	1.50	..	..

Source: World Bank Development Data Platform (DDP), July 6, 2010.
Note: CEMAC = Economic and Monetary Community of Central Africa; EAC = East African Community; ECOWAS = Economic Community of West African States; SADC = Southern African Development Community. .. = negligible.

Table A5.3 *(continued)*

	Personal computers per 100 inhabitants		*Internet users per 100 inhabitants*		*International Internet bandwidth per capita*		*Fixed broadband Internet subscribers*		*Mobile broadband Internet subscribers*		*Total (mobile + fixed)*	
	1998	*2008 (or latest available)*	*1998*	*2008 (or latest available)*	*1998*	*2008 (or latest available)*	*2005*	*2008 (or latest available)*	*2005*	*2009*	*2005*	*2009*
Guinea	0.3	0.5	0.3	0.5	..	0.2	—	—	—	—	—	—
Guinea-Bissau	..	0.2	..	0.2	..	1.3	—	—	—	—	—	—
Kenya	0.3	1.4	0.3	1.4	0.1	21.4	5,399	3,282	—	69,377	5,399	72,659
Lesotho	..	0.3	..	0.3	..	4.9	45	140	—	—	45	140
Liberia	..	..	..	..	..	..	..	..	—	—	—	—
Madagascar	0.2	0.6	0.2	0.6	..	8.1	—	3,488	—	—	—	3,488
Malawi	0.1	0.2	0.1	0.2	0.0	4.6	404	3,400	—	—	404	3,400
Mali	0.1	0.8	0.1	0.8	..	51.5	—	5,272	—	—	—	5,272
Mauritania	0.6	4.5	0.6	4.5	0.1	76.2	164	5,876	—	106,334	164	112,210
Mauritius	8.6	17.6	8.6	17.6	2.5	364.1	5,398	91,734	4,422	168,769	9,820	260,503
Mayotte	..	..	..	..	..	..	..	..	—	—	—	—
Mozambique	0.2	1.4	0.2	1.4	..	3.3	—	10,191	—	92,468	—	102,659
Namibia	2.3	23.9	2.3	23.9	..	26.8	134	320	—	32,211	134	32,531
Niger	0.0	0.1	0.0	0.1	..	11.0	212	617	—	—	212	617
Nigeria	0.5	0.9	0.5	0.9	..	4.7	500	67,776	—	8,756,780	500	8,824,556
Rwanda	..	0.3	..	0.3	..	27.5	1,180	4,241	—	15,177	1,180	19,418
São Tomé and Príncipe	..	3.9	..	3.9	..	50.7	—	760	—	—	—	760
Senegal	1.3	2.2	1.3	2.2	..	237.5	18,028	47,358	—	—	18,028	47,358
Seychelles	12.0	21.2	12.0	21.2	6.5	856.8	948	3,417	—	2,441	948	5,858

Sierra Leone	..	..	..	..	..	..	..	..	—	456	—	456
Somalia	..	0.9	..	0.9	..	0.4	—	—	—	—	—	—
South Africa	5.5	8.5	5.5	8.5	2.4	70.6	165,290	426,000	168,191	5,271,825	333,481	5,697,825
Sudan	0.7	2.0	0.2	10.7	0.0	321.7	1,269	44,625	—	333,843	1,269	378,468
Swaziland	0.2	10.7	..	3.7	..	30.8	—	772	—	—	—	772
Tanzania	..	3.7	0.2	0.9	0.0	2.4	—	6,439	—	601,324	—	607,763
Togo	0.2	0.9	0.6	3.1	0.2	7.6	—	1,911	—	—	—	1,911
Uganda	0.6	3.1	0.2	1.7	0.0	11.7	850	4,798	—	98,105	850	102,903
Zambia	0.2	1.7	0.6	1.1	0.0	7.9	250	5,671	—	—	250	5,671
Zimbabwe	0.6	1.1	1.1	7.6	0.2	9.7	10,185	17,000	—	—	10,185	17,000
Total, all countries							218,564	831,615	172,613	16,395,816	391,177	17,227,431
Percentage of fixed in total											56	5
Economy group												
Low-income, fragile	0.1	0.4	0.2	1.1		1.7		1,656				17,666
Low-income, nonfragile	0.2	1.4	0.2	0.9		16.1		4,950				122,767
Resource-rich	0.5	1.0	0.4	2.3		54.3		45,742				4,973,180
Middle-income	1.4	2.6	1.3	3.6		70.8		104,591				5,035,745
CEMAC	0.2	0.7	0.2	0.7		7.8		395				37,369
EAC	0.2	2.4	0.2	1.2		12.1		4,554				245,480
ECOWAS	0.4	0.9	0.4	1.0		27.5		40,805				4,642,409
SADC	1.3	2.9	1.3	2.7		20.6		83,743				1,207,379
Sub-Saharan Africa	0.3	0.8	0.3	1.0		17.7		22,006				1,029,183

Source: World Bank Development Data Platform (DDP), July 6, 2010.

Note: CEMAC = Economic and Monetary Community of Central Africa; EAC = East African Community; ECOWAS = Economic Community of West African States; SADC = Southern African Development Community. — = data not available; .. = negligible.

Index

Boxes, figures, notes, and tables are indicated by *b, f, n,* and *t,* following the page numbers.

R

S

ECO-AUDIT

Environmental Benefits Statement

The World Bank is committed to preserving endangered forests and natural resources. The Office of the Publisher has chosen to print ***Africa's ICT Infrastructure: Building on the Mobile Revolution*** on recycled paper with 50 percent postconsumer fiber in accordance with the recommended standards for paper usage set by the Green Press Initiative, a nonprofit program supporting publishers in using fiber that is not sourced from endangered forests. For more information, visit www.greenpressinitiative.org.

Saved:
- 9 trees
- 3 million BTUs of total energy
- 854 pounds of net greenhouse gases
- 3,850 gallons of waste water
- 244 pounds of solid waste